Developments in Geotechnical Engineering 3

TUNNELING IN ROCK

Developments in Geotechnical Engineering 3

TUNNELING IN ROCK

by

ERNEST E. WAHLSTROM

Professor of Geological Sciences
Department of Geological Sciences,
University of Colorado, Boulder, Colo., U.S.A.

ELSEVIER SCIENTIFIC PUBLISHING COMPANY
Amsterdam London New York 1973

ELSEVIER SCIENTIFIC PUBLISHING COMPANY
335 Jan van Galenstraat
P.O. Box 330, Amsterdam, The Netherlands

AMERICAN ELSEVIER PUBLISHING COMPANY, INC.
52 Vanderbilt Avenue
New York, New York 10017

Library of Congress Card Number: 72-87968

ISBN 0-444-41064-3

With 121 illustrations and 24 tables.

Printed in The Netherlands

To Kay, loved and esteemed co-worker

PREFACE

Tunnels and other openings excavated in the depths of the earth always have created feelings of excitement tinged with fear in the minds of those who dig them or explore them. The total darkness and the strange silence of an unoccupied tunnel, punctuated at times by the noises of dripping water and sharp adjustments in rocks and supports or the ear-splitting racket and ordered confusion of a high-speed mechanized tunneling operation identify a special kind of environment that is totally unknown and almost incomprehensible to the average person.

A tunnel is a challenge to nature. Construction of an underground opening immediately releases stresses that, over a short or long period of time, tend to close the opening by filling it with rock debris. Some of these stresses are related to presently active dislocations of rocks in the earth's crust, others are residual unbalanced stresses stored in elastic or quasi-elastic rocks, and still others are attributable to the dead weight load of overlying rocks.

When rocks in tunnels are highly incompetent, especially when ground water is present, tunneling becomes a very costly and hazardous operation, and excavation and containment of such rocks present a challenge that requires maximum use of highly technical skill and ingenuity, at times of epic proportions.

The geologist plays an indispensable role in the planning and construction of subterranean openings. He is the professional person who, by training and experience, is most qualified to predict and assess the locations, differing properties, and expected behavior of rocks encountered in excavations below the earth's surface. To fulfill his mission he must meticulously and with wisdom employ all of the field and laboratory instruments and techniques associated with his science and with care and intuition project surface data downward to the level of a planned tunnel or other underground opening. Many times he is frustrated by his inability directly to observe bedrock features at the earth's surface because of extensive cover by soil, alluvium, or vegetation, and he must turn to data from exploratory drill holes and geophysical measurements in an effort to interpret what lies below.

I have written this book from the point of view of a geologist who has had a long and interesting association with the planning and excavation of underground openings, first as a mining geologist and, more recently, as an engineering geologist. My objective is to include within the covers of a small volume a succinct review of the theory and practice of engineering geology as it is applied to tunneling in rocks. It is my hope that the contents of the volume will serve a use as a guide to those who are concerned with geological and related aspects of planning and construction of rock tunnels.

I cannot possibly identify all of those many persons, both geologists and engineers, who have contributed so much to my knowledge and an exciting experience in engineering geology. However, I must, with grateful thanks, single out L.A. Warner of the Department of Geological Sciences of the University of Colorado and V.Q. Hornback, a geologist with the Denver Board of Water Commissioners, who have been highly respected co-workers for many years and who reviewed the manuscript for this book. I wish to acknowledge the kindness of the U.S. Bureau of Reclamation, the Colorado Division of Highways, and the Denver Board of Water Commissioners in providing several photographs used in the text.

Finally, and with great sincerity and loving respect, I thank my wife, Kathryn K. Wahlstrom, for her patient and skillful help in preparing and editing the manuscript.

Ernest E. Wahlstrom

CONTENTS

Preface VII

Chapter 1. About tunnels 1
History of tunneling 1
Some definitions 2
Generalities about tunneling 3
Uses of tunnels 4
Tunnel sections 4
Tunnel supports and lining 7
References 7

Chapter 2. Petrography of unaltered rocks 9
Introduction 9
Rock minerals 10
  General statement 10
  Physical properties of minerals 10
Petrographic methods 14
  Mineralogy of clay minerals 23
Rock fabrics from growth of crystals 26
Fabrics of fragmental rocks 28
  Size classification 28
  Shapes of clasts 29
  Particle orientation 29
  Stratification (bedding) 30
  Matrix and cement 30
Igneous rocks 30
Some shapes of igneous bodies 31
Sedimentary rocks 35
Metamorphic rocks 36
References 37

Chapter 3. Rock alteration 39
Introduction 39
Weathering 40

Mechanical weathering 41
Weathering related to mineral composition 41
Weathering related to ground-water movement 42
Engineering properties of weathered rocks 44
Low-temperature alterations at moderate depths 44
Hydrothermal alteration 44
Silicification 45
Sericitization 46
Argillization 46
Pyritization 46
Propylitization 46
Chloritization 47
Dolomitization 48
Saussuritization 49
Deuteric alteration 49
Pneumatolytic alteration 49
Miscellaneous types of alteration 50

Chapter 4. Elementary rock mechanics 51
Introduction 51
Glossary of definitions 52
Force and stress 53
Mechanical properties of rocks 54
Stress–strain–time relationships 55
Behavior of materials under stress 55
Fractures in rocks 55
Behavior of materials under triaxial stress 57
Effects of temperature on strength of rocks 62
Effects of time on strength of rocks 62
Stresses in underground openings in isotropic elastic rocks 64
Isotropism and anisotropism in rocks 66
Strength of anisotropic rocks 69
Measurement of stress in underground openings 74
Strength of rocks from slope analysis 75
Nonequilibrium forces in rocks 77
Rock bursts 80
Squeezing ground 83
References 84

Chapter 5. Fracture and fold anisotropy in rocks 85
Introduction 85

Deformation of minerals under stress 85
Deformation of elastic and quasi-elastic rocks 87
Deformation of plastic rocks 88
Folds in rocks 88
Engineering aspects of folded rocks 91
Faults 92
Faulting as related to competency of rocks 93
Fault filling and gouge 93
Wall rock alteration associated with faults 96
Major faults 98
Joints 99
Field study and graphical representation of multiple joints 99
Spacing of joints and characteristics of joint surfaces 101
Original joints in igneous rocks 102
Original joints in sedimentary rocks 103
Original joints in metamorphic rocks 104
Tunnel sections in fractured rocks 105
The arching tendency 106
References 108

Chapter 6. Hydrogeology, temperatures, and gases in tunnels 109
Introduction 109
Water in rocks 109
Porosity and permeability 111
Permeability in volcanic rocks 113
Water table and aquifers 113
Sources and temperatures of underground water 114
Fracture permeability 117
Permeability in limestones 118
Squeezing, swelling, and running ground 120
Causes and mechanics of swelling in rocks 121
Swelling pressures of clay-bearing rocks 123
Squeezing ground 125
Running ground 128
Tunneling in squeezing, swelling, and running ground 130
Testing of permeability of rocks 131
Gases in rocks 132
References 133

Chapter 7. Geological investigation of proposed tunnel locations . . . 135
Introduction 135

Geomorphology and subsurface geology 135
Geologic maps and cross-sections of tunnel sites 143
Introduction 143
Kinds of surface geologic maps 147
Geologic cross-sections 150
Columnar sections 151
Geologic and geophysical data from field and laboratory investigations 152
Geophysical investigations 154
Introduction 154
Seismic methods 155
Gravity methods 158
Magnetic methods 158
Electrical methods 159
Remote sensing including aerial photography 163
Side-looking radar imagery (SLAR) 164
Thermal imagery 164
Color infrared photography 165
Core drilling 165
Introduction 165
Core-hole logs 168
Water testing 169
Oriented cores 170
Geophysical investigation and instrumental logging of core holes . . 172
Use of equal-area plots for anisotropic rocks 174
Statistical analysis and computerization of geological data 179
References 179

Chapter 8. Application of geological investigations to tunnel planning and design 181
Introduction 181
Classification of geological materials 182
Feasibility investigations of proposed tunnel sites 190
Design-stage investigations 193
A case history 199
References 202

Chapter 9. Tunneling methods 203
Introduction 203
Conventional tunneling methods 204
Tunneling machines 206
Tunnel supports 208

Ground water in tunnels 213
Conventional versus mole tunneling 216
Access to tunneling operations 218
Tunneling terminology 218
Glossary of tunneling terms 218
References 222

Chapter 10. Geological investigations during tunnel construction . . . 223
Introduction 223
Geological mapping of tunnels 223
Tunnel supports and the geologist 229
Special operations 230
Progress related to geology 232
Final report 235
References 235

Index 237

Chapter 1

ABOUT TUNNELS

HISTORY OF TUNNELING

The beginnings of man's efforts to go underground by means of subterranean excavations are lost in the haze of antiquity, but there is abundant archeological evidence that in Europe Stone Age man sank shafts and drove tunnels to recover flint for the fabrication of sharp-edged implements such as knives, axes, arrowheads, and scrapers. Later, as an elementary knowledge of metallurgy was acquired by primitive people, possibly for the first time in central Asia, underground mining became necessary to supply the increasing demands for metals and alloys in technologically and culturally advancing civilizations. Very early underground excavations for metal-bearing ores have been identified in Caucasia, between the Black and Caspian Seas, and date back to approximately 3500 B.C. From this center and others of unknown location and through ever-expanding communication and trade, knowledge of mining and metallurgical techniques spread in all directions throughout the civilized world.

Many tunnels were built in ancient times by the Babylonians, by the Aztecs and Incas in the search for precious metals, and by the peoples of India, Persia, and Egypt. The Romans, superb planners and engineers, built many tunnels, some connecting with still-standing aqueducts and others as integral parts of highway systems or military operations.

Stone Age man used very primitive tools in underground excavation. Particularly useful to him were picks made of deer antlers, flint axes, and hammers, and wedges made of bone and wood. The discovery of smelting and the production of metals and alloys, first bronze, and later iron and steel, provided materials for increasingly efficient rock excavation. However, even with metal tools and until the first underground use of explosives in the seventeenth and, especially in the eighteenth centuries, mining of hard rocks was a difficult challenge that could be overcome only by fire-setting. For hundreds and perhaps thousands of years underground workings in hard rocks, especially those containing few fractures and fissures, were advanced by building fires against rock faces to cause expansion and spalling. In some operations spalling of the heated rock was accelerated by dowsing it with water. The fractured rock was then separated from the working face with picks, gads, and wedges.

Hand-in-hand with the increasing use of explosives, first "black powder" and later nitroglycerine and dynamite, steel tempering techniques were perfected and permitted efficient and economical hand-drilling of holes for explosives in the hardest of rocks. Near the middle of the nineteenth century steam-powered piston drills, and, later, pneumatic percussion drills, powered by compressed air, made their appearance; and at about the same time, but with less than notable success, several tunneling machines (moles) were invented and given field tests. Without exception, all of the early tunneling machines were dismal failures, but the pneumatic rock drills proved their worth and have played a major role in the development of the modern technology of rock excavation. Again, after more than a century, tunneling machines are being given increasing attention, and, because of many notable successes in rapid and economical driving of tunnels, must be considered as offering a reasonable and competitive alternative to driving tunnels by conventional methods employing pneumatic drills and explosives.

SOME DEFINITIONS

A tunnel is an elongate, narrow, essentially linear excavated underground opening with a length greatly exceeding its width or height. Most tunnels are nearly, or exactly, horizontal, but, for special purposes, tunnels may be driven at angles up to about 30 degrees from the earth's surface. Elongate, linear excavations at angles greater than 30 degrees from the horizontal generally are designated as *shafts,* either *inclined* or *vertical.* A tunnel that provides access from the earth's surface to some underground feature is an *adit.* In mining practice, tunnels following mineralized zones are identified as *drifts,* and tunnels driven in barren rock from the earth's surface to a mineral deposit or from one mineralized area to another are called *crosscut tunnels.*

Tunnels have been driven in a total variety of natural materials, ranging from unconsolidated, water-soaked deposits of clay, sand, gravel, and bowlders to dry, very hard, unfractured rocks. Tunneling in unconsolidated surficial deposits formed by geologically recent deposition on the earth's surface, especially under bodies of water, poses special problems that are not considered in the present treatment. Instead, consideration is given only to tunneling in bedrock. Superficial, unconsolidated natural deposits include wind deposits, stream deposits, deposits in standing bodies of water, glacial deposits and deposits resulting from gravitative movement of soils and rock debris down slopes. Unconsolidated deposits differ from rocks in that there is little or no cohesiveness owing to the formation of a cement or to the development of interlocking crystalline components by one process or another.

Bedrock, on the other hand, has assumed, with geological time, a strength, hardness, and cohesiveness which is evident in a more-or-less firm binding-together of the individual mineral grains and a notable degree of resistance to mechanical disaggregation or chemical decomposition.

GENERALITIES ABOUT TUNNELING

Tunnel construction is one of the most costly and, at the same time, one of the most hazardous of all engineering undertakings. Compare the planning for the construction of a heavy structure such as a dam, a hydroelectric power plant, a skyscraper, a high-rise apartment house, or bridge abutments with the planning for a tunnel hundreds or many thousands of feet in length. Site investigations for most kinds of construction are restricted to relatively small areas at the earth's surface, and, allowing reasonable costs of construction, detailed investigations are not only feasible but indicated. In addition, excavation uncovers unforeseen conditions that can be appraised and, if undesirable, remedied before construction of the designed facility. In contrast, a tunnel *is* the designed facility, and, as construction proceeds, adjustments must be made and construction procedures modified to maintain and protect the opening, as anticipated or unanticipated rock conditions are encountered. Also, because of prohibitive costs, site investigations yielding the detailed quantitative information expected from site investigations of localized heavy structures commonly cannot be made, and recourse must be made to geological and geophysical investigations which, at best, yield predictions based on qualitative or semi-quantitative empiricism and, perhaps, on information from always too few exploratory drill holes. Cost estimates for tunnels prior to construction, especially for long tunnels in regions of geological complexity, are made with great uncertainty because of inability exactly to assess underground conditions. It is not uncommon for an initial cost estimate to be greatly exceeded because of difficulties encountered in sections comprising only a small per cent of the total length of a tunnel.

It is the function of those who plan and design tunnels to make every reasonable effort based on geological, geophysical, and rock-mechanics studies to anticipate undesirable underground conditions and, if possible, to locate tunnels so as to avoid or at least minimize hazardous geological conditions that, from past experience, can be expected to create excessively high costs of tunneling.

The modern technology of underground excavation has reached an advanced stage of development in both theory and practice. Rock-mechanics investigations have provided a vast reservoir of knowledge concerning the behavior of rocks under stress, and this knowledge can be applied to the planning and design of tunnels to the extent that underground geological conditions can be predicted. Unfortunately, underground rock-mechanics studies can be made *only after* the excavation of a tunnel, and serve their most useful immediately practical purpose in determination of the kinds and amounts of supports and tunnel lining that should be installed to prevent short term or long term structural failure of a tunnel.

The ability to anticipate underground geologic conditions is directly related to the simplicity of the geological conditions and the extent to which these conditions are known. For example, tunnels driven in a horizontally-layered massive, well-cemented, relatively unjointed sandstone, through a layer of coal, or through monolithic, relatively unfractured bodies of granite pose no special problems in design and preparation of cost estimates. In contrast, design and cost estimates for tunnels in geologically complex areas, where surface rock exposures may be few in number because of extensive cover, must be based to a large extent on intelligent guesswork, and the planning and design of such tunnels is skilled art rather than quantitative science. It is not surprising that one of the most comprehensive and authoritative treatises on tunneling in recent years is entitled *The Art of Tunnelling* (Széchy, 1967).

USES OF TUNNELS

Tunnels have been constructed for a great variety of purposes. A classification of tunnels by use follows:

(1) Tunnels driven to gain access to economic mineral deposits and to provide haulways for extracted materials. Commonly, such tunnels provide drainage of underground water. In some mining operations tunnels are driven to provide adequate circulation of air in underground workings.

(2) Transportation tunnels, including pedestrian, highway, navigational, and railroad tunnels. These are among the largest, and, at times, the most difficult of all tunnels to excavate.

(3) Water (and sewage) tunnels. These tunnels may or may not be constructed so as to transport liquids under pressure, and a distinction is made between *gravity-flow tunnels* and *pressure tunnels*. The latter are designed to contain without leakage water under a hydrostatic or force-pressure head.

(4) Military tunnels. These tunnels are driven in connection with underground military operations.

(5) Access tunnels to underground chambers or vaults.

(6) Tunnels to provide protection from atomic explosions.

(7) Utility tunnels. Built to contain power and communication transmission lines, gas lines, air-pressure lines, etc.

TUNNEL SECTIONS

Tunnels range in dimensions of cross-sections from those of small galleries driven by painstaking, tedious efforts of miners working with hand-tools to tunnels large enough to accommodate railroad trains, double lanes of highway traffic, or to transport very large volumes of water as in overflow or diversion structures in dams

Fig. 1. Various idealized shapes of tunnel cross-sections excavated in rock prior to installation of supports or lining. The arches (roofs) of many sections are segments of circles (D, E, F, K, L, M, O, P). The arches in G, H, and N are half-ellipses.

and reservoirs. Miners have discovered that a minimum size tunnel that will permit reasonable working space at the heading is one that is at least 7 ft. high and 4 ft. wide, clear of the rock walls or tunnel supports.

Designed shapes and sizes of tunnels in cross-section conform to the planned uses of tunnels and, to some extent, to the nature of the material that it is anticipated will be encountered during excavation. Cross-sectional shapes vary from

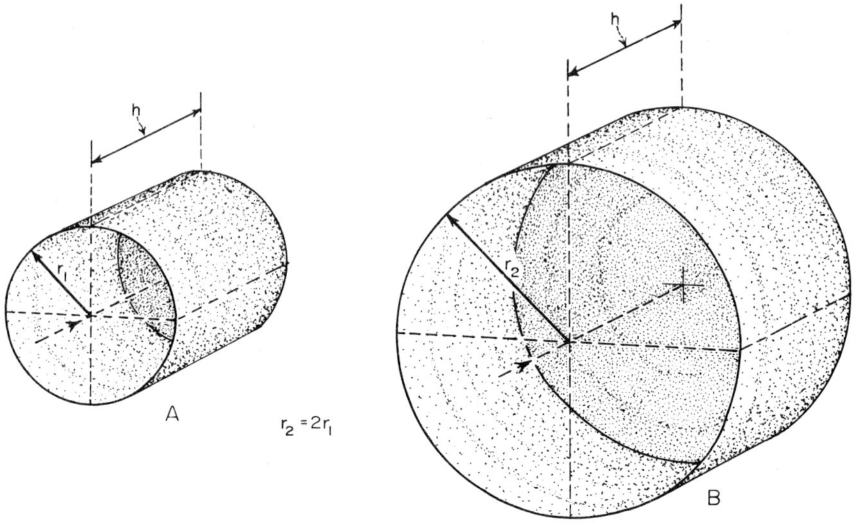

Fig.2. Tunnels of circular cross-section with different radii.

square or rectangular, as, for example, in mining operations in strongly bedded or layered sedimentary rocks, to circular. A common type of cross-section is horseshoe shaped or with the outlines of a gothic-arch to provide maximum stability in the roof portion of the tunnel. Some shapes of cross-sections excavated in rock are shown in Fig. 1.

The area of the cross-section, as well as the shape, is a very important consideration in driving tunnels and to a large extent determines the method of tunneling, especially in incompetent rock requiring extensive support.

The volume of rock that is excavated from a circular tunnel (Fig. 2) is a function of the radius, r, and the length of the tunnel section, h, and is obtained by the formula $\pi r^2 h$. The area of rock exposed in a circular tunnel of length, h, and radius r, including the exposures around the periphery and the exposures in the heading is $2\pi rh + \pi r^2$.

Using these formulas, a comparison is made in Fig. 2 between two tunnel sections of the same length but of different radii. If in Fig. 2A, a radius of 5 ft. and a length of 20 ft. are assumed, the volume of undisturbed rock that has been excavated is 1570.8 cu. ft., and the area exposed in the heading and around the periphery is 706.9 sq. ft. In Fig. 2B the radius is assumed to be 10 ft. Now the volume of excavated rock, because of proportionality to the square of the radius, is 4 times the volume in Fig. 2A, and the exposed area around the periphery and in the heading is 1570.8 sq. ft., 2.2 times greater than in Fig. 2A.

If the rock around the tunnel and in the heading is highly fractured and/or altered and collapses or is squeezed into the tunnel if unconfined by supports, it is clear that the forces causing the collapse or squeezing are effective over a much greater area in the larger tunnel than in the smaller tunnel. Accordingly, the difficulty of providing adequate containment of incompetent rock by installation of supports will be considerably greater in the larger tunnel than in the smaller tunnel. Obviously, in incompetent rock a tunneling method that attacks such rock requires initial borings of small size, with the expectation that such borings can be expanded with careful installation of a growing system of supports. For large tunnel cross-sections in incompetent rock a variety of methods are used in advancing and supporting the tunnel. If the roof has a strong tendency to collapse, the *heading and bench* method is used, whereby the upper part of the tunnel is excavated, supports installed, the lower part of the tunnel is then excavated, and, if necessary, additional supports are installed. Where conditions are extremely adverse, small, heavily supported tunnels are driven and subsequently used to control placement of additional supports as the tunnel is enlarged to full size.

TUNNEL SUPPORTS AND LINING

In mining operations where it is expected that tunnels and other underground workings eventually will be abandoned, no attempt is made to install permanent supports and lining in sections of incompetent rock except, perhaps, in major haulways. The main purpose of installing supports is to provide relatively short-term safety in the excavation. In contrast, in a tunnel that is built to serve a long-time function, such as a traffic tunnel or a water tunnel, efforts must be made during construction to ensure service that will not be interrupted by structural failure within the projected lifetime of the facility.

Until very recent times many tunnels in incompetent rock were supported by heavy timber members which subsequently were removed as masonry walls and arches built from carefully shaped blocks of rock were installed. Modern tunneling makes full use of a variety of supports, including timber, but, especially, steel beams and reinforced concrete.

REFERENCES

Agricola, G. (Georg Bauer), 1556. *De Re Metallica* (in Latin). Translated into English by Herbert C. and L.H. Hoover, *The Mining Magazine,* London, 1912. Reprinted by Dover, New York, 1950.

Murray, D., 1961. *Man Against Earth*. J.B. Lippincott Co. Philadelphia N.J., 176 pp.

Sandström, G.E., 1963a. *History of Tunnelling*. Barrie and Rockcliff, London, 427 pp.

Sandström, G.E., 1963b. *Tunnels*. Holt, Rinehart, and Winston, New York, N.Y., 427 pp.

Széchy, K., 1967. *The Art of Tunnelling.* Akademiai Kiado, Budapest, 891 pp.

Széchy, K., 1969. *Tunnelbau.* Springer, Berlin–Heidelberg–New York, 858 pp.

Chapter 2

PETROGRAPHY OF UNALTERED ROCKS

INTRODUCTION

The physical properties of rocks and the behavior of rocks under stress are expressions of chemical composition and fabric. Wide variations in the properties of rocks correlate exactly with wide variations in mineralogy and fabric. At least an elementary understanding of the classification and various characteristics of rocks is essential in the planning, design, and excavation of underground openings.

This chapter is concerned with the composition and physical characteristics of unaltered (fresh) rocks as they are measured or estimated in hand-specimen-size samples, either in the field or in the laboratory. Techniques of examination utilize a broad spectrum of methods ranging from simple visual examination with or without the aid of a hand lens to comprehensive examination with all or most of the great variety of tools available in a modern petrographic laboratory.

Two aspects of rocks are of prime importance in classification: (1) the mineral composition and (2) the fabric. Fabric is an expression of the size or sizes in mineral aggregates of mineral grains or particles, the shapes of the mineral grains, their orientation, and their manner of articulation. There are two broad categories of fabrics based on the natural histories of rocks: (1) fabrics which are the result of accumulation of fragments of minerals and rocks of a wide range of sizes followed by lithification, and, in many instances, development of a cement, and (2) fabrics resulting from crystallization from molten igneous magmas or resulting from more or less pervasive recrystallization of existing rocks. A broad distinction may be made between rocks which have developed by the accumulation of particles by sedimentary processes, so-called *accumulates*, and rocks which have formed by crystallization or recrystallization of preexisting liquids or solids, so-called *crystallates.*

Considering the origin of rocks, a genetic distinction may be made among (1) *igneous rocks,* which formed by cooling and solidification of fluid, molten magmas, (2) *sedimentary rocks* which were deposited from fluid suspensions or solutions to build up stratified accumulations, and (3) *metamorphic rocks,* which, in response to elevated temperatures and/or pressures in the earth's crust, were transformed by changes in fabric and mineralogy into rocks differing more or less from the original rocks. All gradations exist among the three classes of rocks, and classification of a rock as belonging to one group or another at times must be made on the basis of arbitrarily defined criteria.

This chapter is concerned with the systematics of the petrography of unaltered rocks, that is, rocks which have not been changed as a consequence of weathering at the earth's surface or have not undergone notable changes in mineralogy and/or fabric by moderately warm to hot, circulating underground gases or aqueous solutions. The particular changes in rocks resulting from penetration by surface waters or warm to hot solutions of deep-seated origin are described in Chapter 3, the following chapter.

ROCK MINERALS

General statement

Rock minerals are the abundant minerals constituting the bulk of the great variety of rock types exposed at the earth's surface. Most rocks have simple mineralogies and rarely contain more than five or six volumetrically significant mineral components, and, not uncommonly, only one or two. A broad distinction is made between the *silicates,* the dominant constituents of most rocks, and the *non-silicates,* present in the earth's crust in considerably smaller amounts.

Mineralogists classify minerals on the basis of chemical composition and crystal structure. An outline classification of important rock minerals is given in Table I. Excluded from Table I are minerals which commonly are present in subsidiary amounts in most natural occurrences and rarely are volumetrically important. Excluded also are the metallic minerals found only in concentrations in economic ore deposits.

Physical properties of minerals

Rock minerals in grains large enough to be examined with the unaided eye or with a hand lens reveal properties that are useful in identification. Very small mineral particles require examination under a microscope or by other means before identification is possible. Following is a discussion of properties that can be identified in mineral grains about a millimeter or more in diameter.

Properties depending on reflection and absorption of light

Visible light incident on minerals is partly reflected and partly transmitted and more or less absorbed, and, depending on the nature of reflection and transmission and/or absorption, a broad distinction is made between substances with *nonmetallic lusters* (such as quartz) and substances with *metallic* and *submetallic lusters* (such as magnetite and ilmenite). Most rock minerals have nonmetallic lusters, which can be further characterized by such descriptive terms as *vitreous* or *glassy, resinous, pearly, greasy, silky,* and *adamantine.*

TABLE I

Summary classification of important rock minerals

Silicates	*Nonsilicates*
Orthosilicates and ring silicates	oxides (including magnetite, hematite, and ilmenite)
olivine	hydroxides (especially limonite)
garnet	sulfides (especially pyrite and marcasite)
sillimanite, andalusite, kyanite	sulfates (especially barite, gypsum, and anhydrite)
staurolite	carbonates (including magnesite, calcite, dolomite and siderite)
epidote, zoisite, clinozoisite	halides (fluorite and halite)
melilite	
cordierite	
tourmaline	
Chain silicates	
pyroxene	
amphibole (including hornblende and tremolite)	
Sheet silicates	
mica group (including biotite and muscovite)	
talc	
chlorite	
serpentine	
clay minerals	
Framework silicates	
feldspar group (including potassium feldspars and plagioclase)	
quartz, chalcedony, opal	
feldspathoids (especially nepheline and leucite)	
zeolites	

Color results from reflection or transmission of certain frequencies of the visible spectrum and more or less selective absorption of others.

Hardness

Hardness measures the resistance to scratching, or indentation under pressure by an object such as a sphere or a diamond point. Accordingly, a distinction is made

TABLE II

Mohs' scale of hardness

1. talc	6. orthoclase
2. gypsum	7. quartz
3. calcite	8. topaz
4. fluorite	9. corundum
5. apatite	10. diamond

between *scratch hardness,* which is most useful for the present purposes, and *indentation hardness.* Scratch hardness is determined in relationship to *Moh's scale* (Table II). Each mineral in the scale will scratch all minerals of lower numerical hardness in the scale. Compared with the Mohs' scale are some hardnesses useful in hardness tests: fingernail, 2 plus; copper, about 3; blade of pocket knife, 5 plus; window glass, 5½; and file-steel, 6½. With practice, the hardness of a substance less than 5½ is easily estimated using only a knife blade.

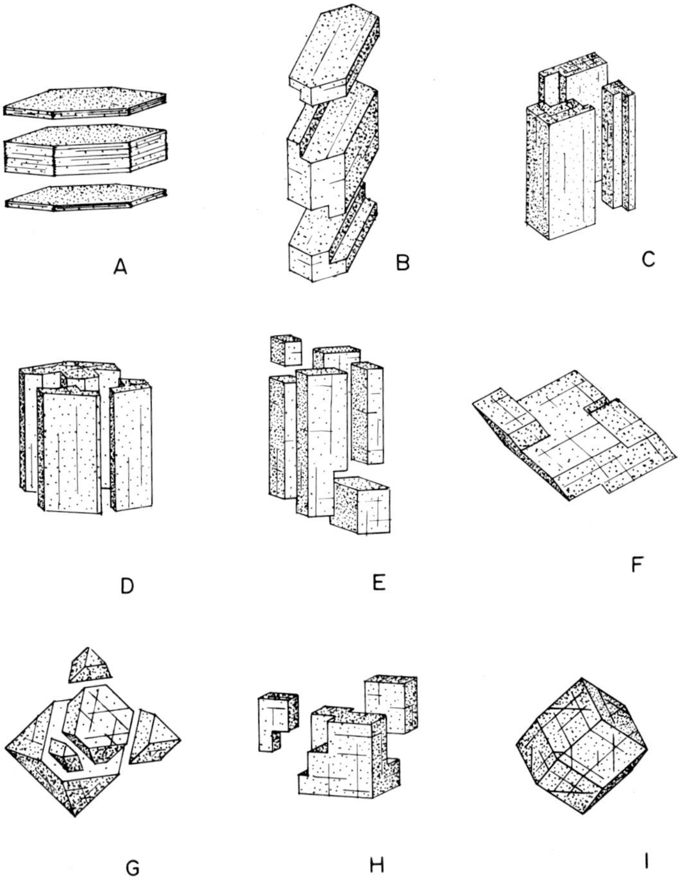

Fig. 3. Several kinds of cleavage. A. Perfect cleavage in mica. B. Two cleavages nearly at right angles in a feldspar crystal. One cleavage (basal) is excellent and the other is good. C. Two directions of equally excellent cleavage in pryoxene. Cleavages intersect nearly at right angles. D. Three directions of excellent cleavages intersecting at 60° in a prismatic hexagonal crystal. E. Two directions of excellent cleavage and one direction of good cleavage, all intersecting at right angles. F. Three directions of equally perfect cleavage in calcite. G. Four directions of equally excellent cleavage in fluorite. H. Three directions of equally perfect cleavage intersecting at right angles in halite (rock salt). I. Six directions of equal cleavage parallel to faces of a dodecahedral crystal.

Cleavage

Cleavage is a constant property of a mineral substance which is revealed by splitting on impact in one or more directions. The quality of an observed cleavage may be characterized as poor, fair, good, excellent, or perfect, depending on the ease of splitting and the perfection of the cleavage surface. Some substances, for example quartz, have no cleavage; in contrast, mica has a single, perfect cleavage. Fig. 3 illustrates several kinds of cleavages and combinations of cleavages.

Fracture

Fracture defines the manner in which a mineral breaks in any direction not a cleavage direction. Several kinds of fracture are *uneven* or *irregular, hackly, fibrous* or *splintery,* and *conchoidal* (like a clam shell).

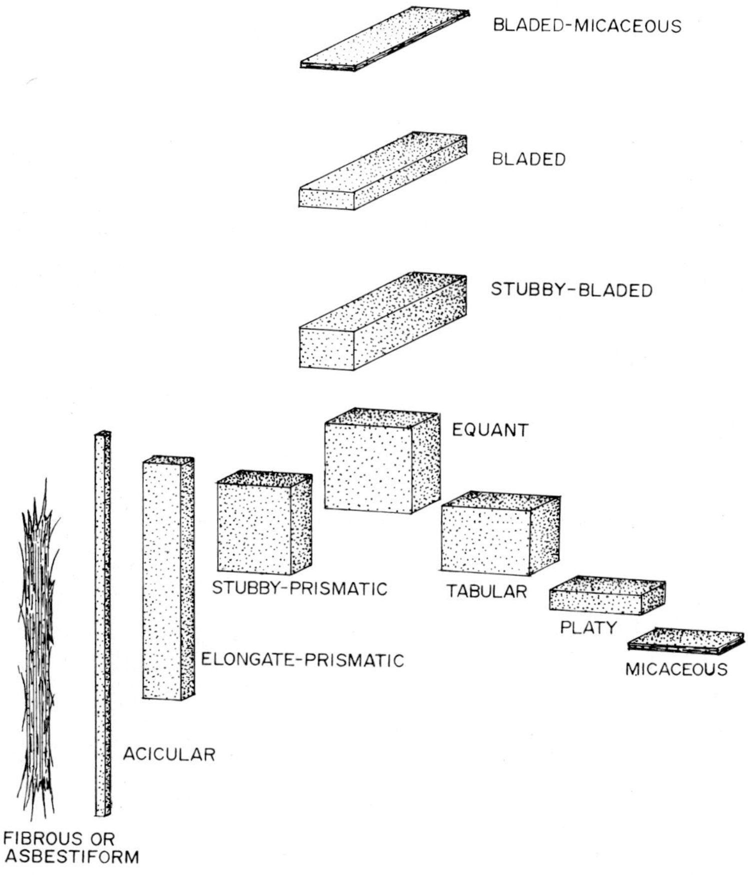

Fig. 4. Various kinds of habits (shapes) assumed by minerals.

Habit

Habit describes the shape assumed by a mineral during crystallization. Completely formed crystals, bounded on all sides by crystal faces, are rare in rocks, but grains may develop some faces or none at all. Nevertheless, characteristic shapes often develop and are useful in identification of minerals. Some habits of minerals are indicated diagrammatically in Fig. 4.

Twinning

Crystals, either with growth faces or without, are a consequence of growth by building up of a systematic arrangement of atoms or molecules. The particular arrangement defines the *structure* of a crystal and is to a major extent dependent upon the chemical composition. Many crystals grow as *twins,* in which adjacent segments appear to be rotated with respect to each other. On either side of the contact between the segments the crystal structures are not parallel, but bear a symmetrical relationship to each other. Such crystals are described as *twinned crystals,* and in an individual crystal there may be two, or multiple segments.

Specific gravity and heft

Specific gravity is defined by the relationship:

$$\text{Specific gravity} = \frac{W_a}{W_a - W_w} \qquad (2\text{-}1)$$

where W_a is the weight of a test sample in air and W_w is the weight of the sample when immersed in water. Specific gravity is the ratio of the weight of a given volume of a substance to the weight of an equal volume of water, and generally is measured with a specific-gravity balance.

Heft is the feel of weight of an object held in the hand and permits qualitative distinction between substances of high specific gravity and those of moderate to low specific gravity. For example barite is a nonmetallic substance with an unusually high heft (Sp. G. = 4.5, compared with quartz, 2.65).

Megascopic properties of rock minerals

The properties of the volumetrically important rock minerals, as they are observed in hand-specimen with or without a hand lens, are summarized in Table III. The minerals are listed in order of increasing hardness.

PETROGRAPHIC METHODS

The petrographer working in a modern laboratory has available to him a wide variety of instruments and techniques for the determination of the chemical com-

TABLE III

Properties of abundant rock-making minerals (in order of increasing hardness)

Name and composition	*Hardness*	*Cleavage(s)*	*Habit*	*Luster and color*	*Specific gravity*	*Remarks*
Talc Mg-silicate with (OH)	1	one perfect	platy, micaceous	nonmetallic; pearly, greasy; colorless, pale colors	2.7–2.8	greasy feel
Hematite, earthy Fe_2O_3 + impurities	1+	–	microcrystalline	metallic to submetallic; red to vermillion	5±	red iron ore
Limonite, earthy $HFeO_2$ + impurities and water	1+	–	microcrystalline to amorphous	metallic to submetallic; yellow, brown	5±	weathering of iron-bearing minerals
Gypsum (selenite) $CaSO_4 \cdot 2H_2O$	2	one perfect; two poor, commonly obscure	tabular	nonmetallic; vitreous, pearly or silky; colorless or pale colors	2.32	scratched by fingernail
Mica, muscovite, white mica K–Al-silicate with (OH)	2–2½	one perfect	micaceous, tabular	nonmetallic; vitreous	2.76–3.1	flexible and elastic
Chlorite complex Mg–Fe–Al silicate with (OH)	2–2½	one perfect	micaceous, tabular	nonmetallic; vitreous to dull; especially green, but also brown	2.6–2.9	flexible but not elastic
Clay minerals Al-silicates with (OH) and Mg–Al-silicates with (OH)	2–2½	microcrystalline; one perfect cleavage	earthy, micaceous	nonmetallic, dull; white and pale colors	2.6±	described elsewhere

TABLE III (continued)

Name and composition	*Hardness*	*Cleavage(s)*	*Habit*	*Luster and color*	*Specific gravity*	*Remarks*
Serpentine Mg-silicate with some Fe and (OH)	2–5	one perfect, not generally seen because of small grain size	microcrystalline, micaceous	nonmetallic, greasy; green, gray	2.5–2.6	sometimes asbestiform
Halite (rock salt) NaCl	2½	excellent, three directions at right angles	equant	nonmetallic; vitreous, greasy; colorless, pale colors	2.16	dissolves in water
Mica, biotite, black mica K–Mg–Fe–Al-silicate with (OH)	2½–3	one direction, perfect	platy, micaceous	nonmetallic; vitreous; black, dark green, or brown	2.8–3.2	flexible and elastic, or brittle
Calcite (calcspar) $CaCO_3$	3	three directions, excellent, at acute angles	acicular to tabular	nonmetallic, vitreous; colorless	2.71	dissolves with effervescence in dilute HCl
Anhydrite $CaSO_4$	3–3½	one perfect, one less perfect, one good	tabular, prismatic, equant	nonmetallic, vitreous; colorless, pale colors, esp. blue or violet	2.89–2.98	hydrates to gypsum
Barite $BaSO_4$	3–3½	one perfect, two less perfect	tabular	nonmetallic, vitreous to pearly; colorless white, pale colors	4.5	high specific gravity
Dolomite $CaMg(CO_3)_2$	3½–4	three excellent at acute angles	equant	nonmetallic, vitreous; colorless, pale colors	2.85	soluble in warm, dilute HCl
Siderite $FeCO_3$	3½–4	three excellent at acute angles	equant	nonmetallic, vitreous or dull; light to dark brown	3.9±	in iron-rich sedimentary rocks

Magnesite $MgCO_3$	3½–5	three good at acute angles	equant	nonmetallic, vitreous or dull; white or faint colors	3.0–3.2	in magnesian rocks
Zeolites Ca–Na–Al-silicates with (OH) and H_2O	3½–5½	variable	equant to acicular	nonmetallic, vitreous or dull; colorless, white, faint colors	2.0–2.4	complex group of minerals
Fluorite CaF_2	4	excellent in four directions	equant	nonmetallic, vitreous; colorless, or pale colors, esp. lavender	3.18	relatively rare
Melilite Ca–Mg–Al-silicate	5	one direction, poor; one direction generally indistinct	tabular to stubby prismatic	nonmetallic, vitreous to resinous; white, pale colors	2.9–3.1	relatively rare
Limonite, compact $HFeO_2$ + water and impurities	5–5½	not observable	amorphous or cryptocrystalline	submetallic; brown, reddish brown	4.3±	powder is yellowish brown
Pyroxene, augite, diopside Ca–Mg–Fe–Al-silicate and Ca–Mg–Fe-silicate	5–6	two directions good, nearly at right angles	stubby, prismatic	nonmetallic, vitreous; pale colors to dark green, brown or black	3.2–3.4	very common
Pyroxene, enstatite, bronzite, hypersthene Mg–Fe-silicates	5–6	two directions good, nearly at right angles	stubby, prismatic to acicular	nonmetallic, vitreous; pale colors to dark green or brown	3.3±	Mg-rich rocks
Amphibole, hornblende Ca–Na–Mg–Fe–Al-silicate with (OH)	5–6	two directions, excellent at 55 and 125 degrees	prismatic, asbestiform, acicular	nonmetallic, vitreous; black, dark brown or green	3.2±	very common

TABLE III (continued)

Name and composition	*Hardness*	*Cleavage(s)*	*Habit*	*Luster and color*	*Specific gravity*	*Remarks*
Amphibole, tremolite Ca–Mg-silicate with (OH)	5–6	two directions excellent at 55 and 125 degrees	prismatic, asbestiform, acicular	nonmetallic, vitreous; colorless, or pale colors	3.0–3.3	in metamorphosed limestones
Opal SiO_2 plus some H_2O	5–6	none	amorphous	nonmetallic, vitreous; dull; white and pale colors	1.9–2.2	low specific gravity
Feldspathoid (nepheline) Na–K–Al-silicate	5–6	three directions distinct at 60 degrees	equant, stubby prismatic	nonmetallic, greasy, vitreous; white, gray, pale colors	2.60–2.65	gelatinizes in HCl
Kyanite Al_2SiO_5	5–7	one perfect	prismatic to elongate tabular, bladed	nonmetallic, vitreous to pearly; commonly colorless or blue	3.55–3.66	metamorphic rocks only
Hematite, crystalline Fe_2O_3	5½–6	none	tabular, scaly	metallic; black to reddish brown	5.26	powder red; nonmagnetic
Ilmenite $FeTiO_3$	5½–6	none	platy	metallic to submetallic; iron-black	4.7	slightly magnetic; powder brownish red
Feldspathoid, leucite K–Al-silicate	5½–6	none	equant	nonmetallic, vitreous; white, gray	2.47	fairly rare
Potassium feldspar (orthoclase, microcline, sanidine)	6	one excellent, one good, nearly or exactly at right angles	equant to stubby prismatic	nonmetallic, vitreous; colorless, pale colors	2.57±	very common

Plagioclase, sodic (albite) Na ≫ Ca	6	one excellent, one good, nearly at right angles	equant to tabular	nonmetallic, vitreous; colorless, white	2.62	feldspar
Plagioclase, intermediate (Oligoclase, andesine) Na–Ca–Al-silicate Na > Ca or Na ≅ Ca	6	one excellent, one good, nearly at right angles	equant to tabular	nonmetallic, vitreous; white, gray	2.65–2.69	feldspar
Plagioclase, calcic (Labradorite, bytownite, anorthite) Ca–Na–Al-silicate Ca ≅ Na or Ca >> Na	6	one excellent, one good, nearly at right angles	tabular	nonmetallic, vitreous; gray	2.71–2.74	feldspar
Magnetite, magnetic iron ore Fe_3O_4	6	none	equant	metallic; iron-black	5.18	magnetic
Pyrite FeS_2	6–6½	none	equant, commonly in cubes	metallic; yellow or yellowish white	5.02	common sulfide mineral
Marcasite FeS_2	6–6½	none	prismatic or tabular	metallic; yellowish white	4.89	commonly in radiating aggregates
Zoisite and clinozoisite Ca–Al-silicate with (OH)	6–6½	one perfect, one imperfect, at obtuse angle	tabular, elongate	nonmetallic, vitreous; white, pale green, colorless, gray	3.35	common secondary mineral
Sillimanite (fibrolite) Al_2SiO_5	6–7	one perfect	prismatic, acicular	nonmetallic, vitreous or dull; white gray, pale colors	3.23	needles common; metamorphic rocks

TABLE III (continued)

Name and composition	*Hardness*	*Cleavage(s)*	*Habit*	*Luster and color*	*Specific gravity*	*Remarks*
Epidote Ca–Al–Fe-silicate with (OH)	6–7	one perfect, one good, at obtuse angle	tabular, elongate	nonmetallic, vitreous; green to black	3.35–3.45	common secondary mineral
Olivine Mg–Fe-silicate	6½–7	none apparent	tabular, prismatic	nonmetallic, vitreous; gray-green, brown	3.27–4.37	conchoidal fracture
Garnet Ca–Mg–Fe–Al-silicate	6½–7½	none	equant	nonmetallic, vitreous or dull; colorless (rare), red, brown, green, black	3.5–4.3	commonly in crystals
Chalcedony (chert) SiO_2	6½–7	none	microcrystalline	nonmetallic, dull, waxy	2.6	conchoidal fracture
Quartz SiO_2	7	none	equant to prismatic; six-sided prisms	nonmetallic, vitreous; colorless, white, gray, pale colors	2.67	very common
Staurolite Fe–Al-silicate with (OH)	7–7½	not apparent	prismatic	nonmetallic, vitreous; resinous; black, green, brown	3.65–3.75	metamorphic rocks; crystals common
Cordierite Mg–Fe–Al-silicate	7–7½	one direction, poor	stubby prismatic	nonmetallic, vitreous, dull; colorless and pale colors	2.60–2.66	metamorphic rocks
Tourmaline Complex. Ca–Na–Mg–Fe–Al-silicate with (F) and (OH)	7–7½	none	prismatic to acicular; trigonal cross-sections	nonmetallic, vitreous; pale colors (rare) black, brown, green	3.0–3.25	pegmatites and metamorphic rocks
Andalusite Al_2SiO_5	7½	two directions good, nearly at right angles	prismatic	nonmetallic, vitreous; dull; white, gray pale colors	3.16–3.20	metamorphic rocks only

position and the identification of minerals by destructive and nondestructive methods. However, one of the most rewarding methods for the study of mineral aggregates in rocks utilizes thin sections about 0.03 mm. thick mounted between a glass slide and a cover glass. Thin sections permit detailed examination of the mineralogy and fabric of rocks, and particular use is made of a polarizing microscope. Generally the examination is made in transmitted light, but in rocks containing opaque minerals the technique is modified to permit examination in polarized reflected light.

The theory of the passage of light through crystals is complex (Wahlstrom, 1969) and identification and determination of the composition of minerals under the polarizing microscope requires considerable training and skill (Heinrich, 1965; Moorhouse, 1959). Procedures consist of measuring optical properties which are characteristic and diagnostic for each mineral substance.

Fig. 5. Hand specimens and thin sections of several rocks. A. Granite (quartz monzonite) porphyry. Igneous. B. Anorthosite. Igneous. C. Micaceous sandstone (arkose). Sedimentary. D. Garnetiferous mica schist. Metamorphic.

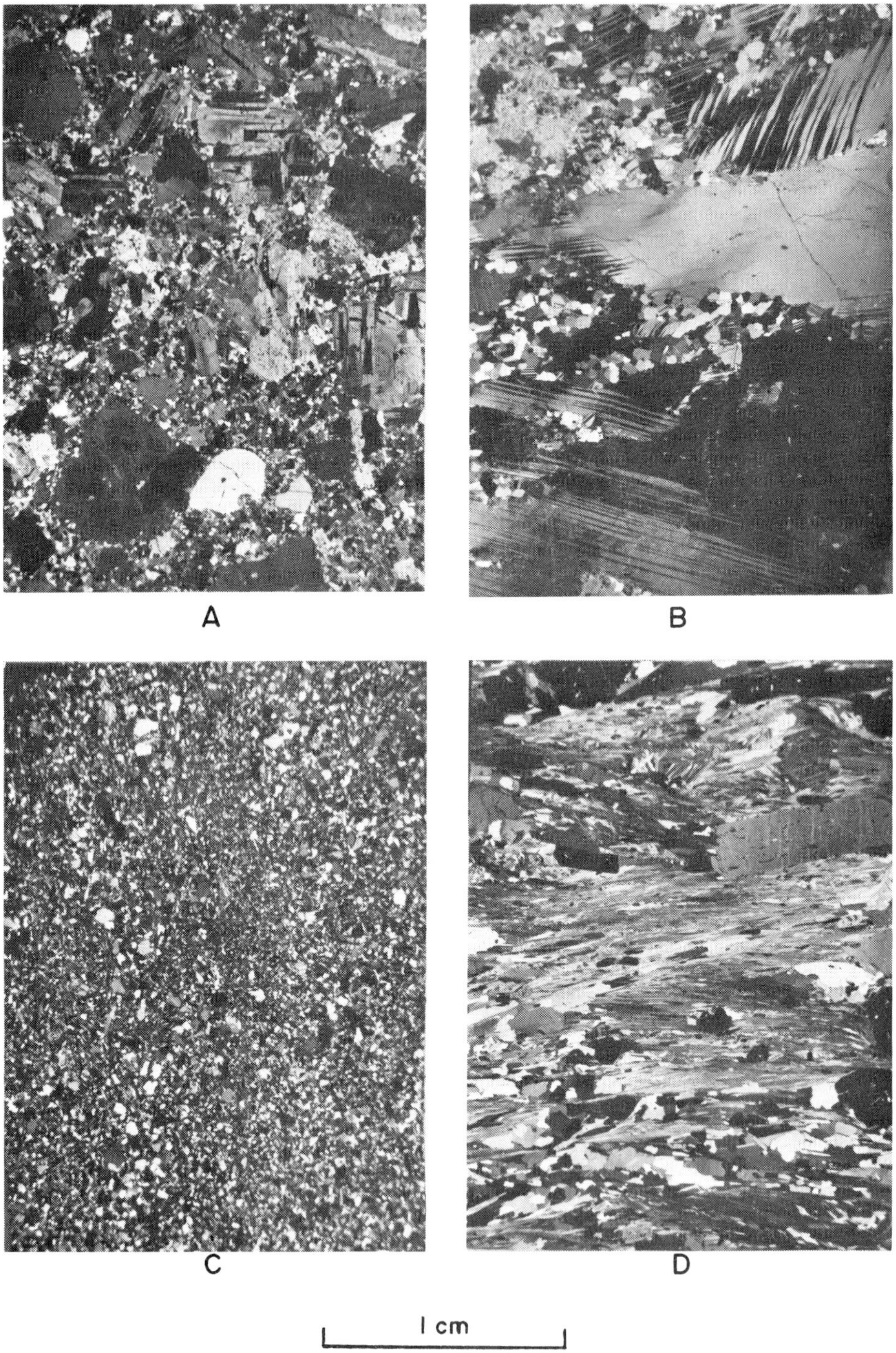
A
B
C
D
1 cm

However, even the novice can learn much about the fabric of a rock, including observation of grain sizes, shapes of mineral grains, manner of articulation, etc., by examining thin sections in polarized light.

Fig. 5 shows hand specimens and thin sections of several rock types. Fig. 6 shows photomicrographs of the thin sections in Fig. 5 viewed in polarized light, and clearly brings out the mineralogy and fabric of each sample.

Thin-section study is particularly effective in the study of rocks consisting of mineral grains of sufficient size to enable measurement of optical properties. Such rocks have grain sizes that can be described as *microcrystalline*, or if grains are visible in hand specimen, as *megacrystalline*. Many rocks, if crystalline, are so fine-grained that the individual grains are at or below the limit of visibility under the microscope and are said to be *cryptocrystalline*. Examples are shale, claystone, slate, and exceedingly fine-grained igneous rocks such as rhyolite. In addition, there are some igneous rocks that are glassy, and some sedimentary rocks contain abundant amorphous, noncrystalline materials. Obviously the polarizing microscope is of slight use in the identification of cryptocrystalline or amorphous components of rocks, but it enables the determination of their presence so that appropriate special techniques may be planned in investigations.

In cryptocrystalline or glassy igneous rocks chemical analyses are required for establishment of a rock name. However, for the study of the important group of rocks consisting partly or entirely of clay minerals a very comprehensive technology has been developed.

Mineralogy of clay minerals

The term *clay mineral* is used in a twofold sense: it is applied to a mineralogically well-defined group of minerals, and, in addition, it is applied to natural substances of very small grain size, of the order of 5 microns or less, found in clay deposits or soils.

The mineralogy of clay minerals is complex, but in a simple classification three important groups may be identified:

(1) *Kaolin group.* Clay minerals chemically identified as Al-silicates containing hydroxyl.

Fig. 6. Photomicrographs in polarized light of thin sections in Fig. 5. A. Granite (quartz monzonite) porphyry containing larger crystals of quartz, orthoclase, and plagioclase in a finer grained aggregate of quartz and feldspar (ground mass). B. Anorthosite consisting dominantly of calcic plagioclase (labradorite). Note crushed aspect of rock. C. Micaceous sandstone (arkose) consisting of angular grains of quartz and feldspar. D. Section at right angles to layering in a mica schist. Plates of mica (muscovite) alternate with layers rich in feldspar crystals. Present also are garnet crystals (black) and a crystal of staurolite (upper right).

(2) *Montmorillonite group.* Clay minerals of complex composition usually containing magnesium, iron, aluminum, silica, and water. Minerals in this group tend to swell on contact with water.

(3) *Illite group.* Clay minerals containing potassium or sodium, aluminum, silica, and hydroxyl.

In addition, many clay aggregates contain micron-size particles of chlorite, and/or muscovite (sericite) and, less commonly a greenish micaceous mineral sometimes formed by the chemical degradation of biotite mica and called *glauconite.* Vermiculite, an altered biotite mica which expands notably when heated, also is classified by some writers as a clay mineral.

The clay minerals, with rare exceptions, are sheet silicates forming tiny micaceous plates that have many characteristics in common with the true micas. Some clay particles consist of alternating layers of different clay minerals and are described as *mixed-layered-silicates.* A familiar example is mixed-layer illite–montmorillonite.

Owing to the very tiny size of the particles and large surface forces of attraction clay minerals have a strong affinity for water, and aggregates tend to swell and become dispersed in contact with water. Members of the montmorillonite group not only attract water to the surfaces of particles but also incorporate water within the bodies of crystals with notable concomitant expansion and development of swelling pressures. An abundant natural material containing montmorillonite is *bentonite,* a highly altered volcanic ash which, starting with the dry state, expands to a volume 20–30 times the original volume on contact with water.

Three techniques are outstanding in the identification and classification of the clay minerals. In order of decreasing utility they are: (1) the X-ray diffractometer method, (2) differential thermal analysis, and (3) examination under an electron microscope.

An X-ray diffractometer is constructed so that a powdered mineral sample mounted on a glass slide can be rotated at a fixed angular rate in an X-ray beam. Simultaneously, and rotating about the same axis a pick-up device, such as a Geiger tube, detects diffracted beams from the mineral sample at critical angles of reinforcement according to Bragg's equation and transmits the impulse in the pick-up device to a strip-chart recorder. The pick-up device rotates at twice the rate of the sample and for a rotation of θ to obtain critical reinforcement of the diffracted beams the pick-up device has rotated through an angle of 2θ.

Spacing of crystal planes then can be calculated from peaks on the strip-chart recording by using the Bragg equation:

$$n\lambda = 2d \sin \theta \qquad (2\text{-}2)$$

where λ is the wave length of the X-radiation and d is the spacing of a particular set of crystal planes.

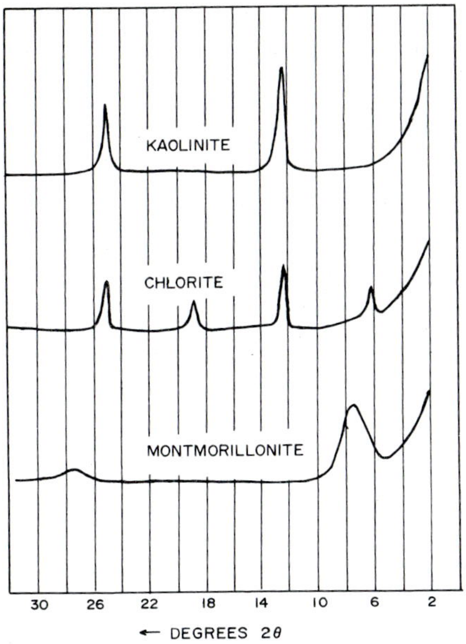

Fig. 7. Idealized X-ray diffractograms for some mineral components of clays.

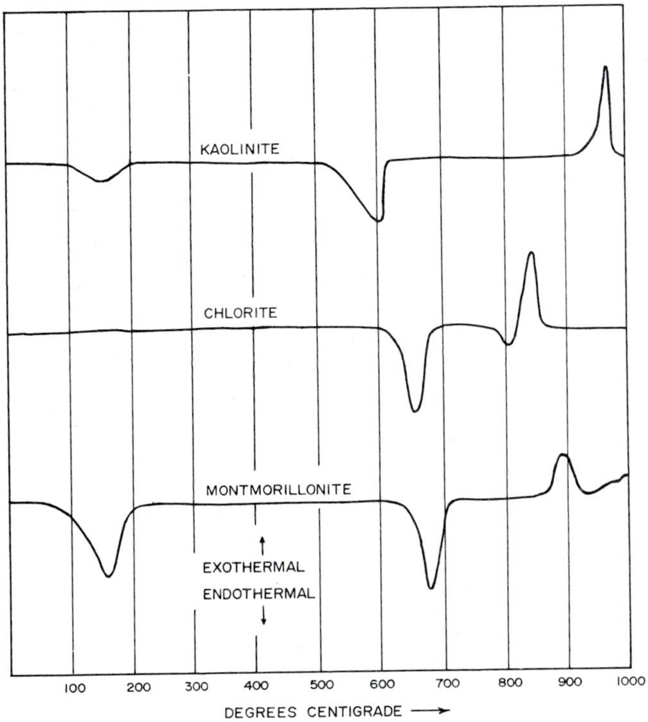

Fig. 8. Differential thermal curves for some mineral components of clays.

Fig. 7 illustrates idealized X-ray diffractograms for some clay minerals. In practice the clay samples are examined before and after chemical and heat treatment, enabling identification of structural changes characteristic of each substance.

The differential thermal technique enables determination of the temperatures and magnitudes of exothermic and endothermic changes in samples as they are heated at a uniform rate in a differential thermal analyzer. By means of thermocouples embedded in the mineral sample and in a thermally inert substance such as aluminum oxide changes in the composition of the mineral, such as loss of water, or changes in crystal structure can be identified in a recording on a strip-chart. Fig. 8 illustrates idealized differential thermal curves for three clay minerals.

The electron microscope is useful in ascertaining sizes and shapes of clay minerals and reveals characteristic habits for each species of clay mineral.

ROCK FABRICS FROM GROWTH OF CRYSTALS

In the study of rocks the term *crystal* refers to a mineral grain which has a crystalline structure without reference to whether crystal faces have developed or not. A distinction may be made among crystals which are *anhedral,* showing no crystal faces, *subhedral,* showing partial development of crystal faces, and *euhedral,* exhibiting a complete development of faces.

During growth some crystals exhibit a strong tendency to develop crystal faces and are said to have a high *form energy.* Other substances have low to moderate form energies, and in the competition for space in a crystallizing aggregate can develop only anhedral or subhedral crystals.

TABLE IV

Grain-size classification for crystalline rocks

Rocks	*Grain-size* (mm)
Phaneric	
very coarse-grained macrocrystalline	> 4.0
coarse-grained macrocrystalline	4.0–1.0
coarse-grained mesocrystalline	1.0–0.5
medium-grained mesocrystalline	0.5–2.5
fine-grained mesocrystalline	0.25–0.10
very fine-grained mesocrystalline	0.10–0.05
Aphanic	
finely-crystalline aphanic	0.05–0.01
microcrystalline aphanic	0.01–0.001
cryptocrystalline aphanic	< 0.001

Fabrics in crystalline rocks originate in a variety of ways, but in a broad sense primarily (1) by direct crystallization from solutions in open spaces or, as in igneous rocks, by crystallization from molten magma, (2) by crystallization or recrystallization of solid materials in existing rocks, and (3) by introduction of materials from an outside source, resulting in partial to complete replacement of existing

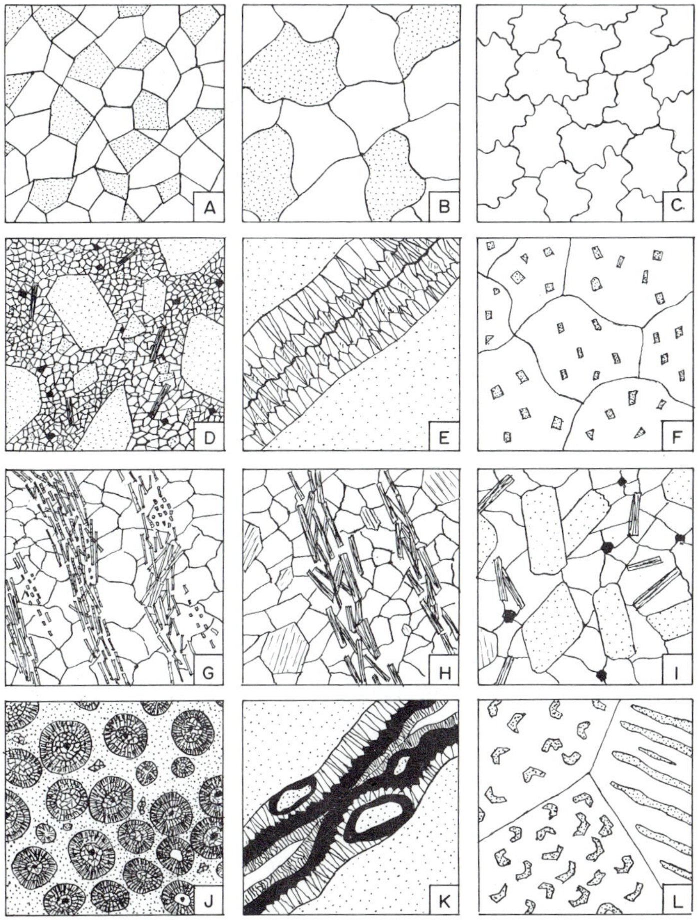

Fig. 9. Some fabrics in rocks. No scale. A. Equigranular-mosaic fabric. B. Equigranular-mutual fabric. C. Equigranular-sutured fabric. D. Porphyritic fabric. Larger crystals (phenocrysts) are embedded in a finer grained matrix (groundmass). E. Vein displaying crustified, cockade fabric. F. Poikilitic fabric. Small crystals are enclosed by larger anhedral crystals. G. Acicular and granular fabrics in a gneiss. H. Platy (micaceous) and equigranular fabric in a gneiss. I. Hypautomorphic-granular fabric. Mineral grains show euhedral, subhedral, and anhedral outlines. J. Oolitic fabric. K. Vein showing cockade fabric alternating with colloform fabric. Black is amorphous to cryptocrystalline material. L. Graphic fabric. Rods of quartz penetrate anhedral crystals of feldspar.

materials. A further distinction may be made between *primary* crystalline fabrics and *secondary* crystalline fabrics, depending on the time of development.

Examples of primary crystalline fabrics are found in igneous rocks and open cavity fillings. Secondary crystalline fabrics are seen in metamorphic rocks which have formed by recrystallization of preexisting rocks, in sedimentary rocks which have recrystallized during or subsequent to lithification, and in replacement deposits formed by introduction of materials in solution and by diffusion into rocks.

For equigranular crystalline rocks the grain-size classification in Table IV is useful. Average diameters are assumed. For rocks containing more than one size of mineral grain the average size is estimated or the actual dimensions of the grains are given.

Several kinds of fabrics in crystalline rocks are shown in Fig. 9. Considering that fabric is an expression not only of (1) grain size, but also of (2) crystal shape (habit), and (3) the manner of articulation, it is understood that a very large number of combinations of the three elements of fabric can exist in rocks.

From an engineering point of view originally crystalline igneous rocks or other rocks which have crystallized or recrystallized from an original nonliquid condition are among the most competent of all known natural materials.

FABRICS OF FRAGMENTAL ROCKS

Fragmental rocks consist dominantly or entirely of mineral or rock fragments. The fragments are sometimes called *clasts* and, accordingly, comprise the so-called *clastic* rocks. Accumulations of mineral and rock fragments associated with more or less explosive volcanic activity are identified as *pyroclastic deposits.* Most sedimentary rocks, including pyroclastic rocks, are initially clastic, and result from deposition from suspensions or traction-carried loads of particles transported water, air, or ice, or accumulations of materials moving downslope under the influence of gravity either on the earth's surface or beneath water. Sedimentary rocks consisting of minerals precipitated from solutions such as salt crystals or calcareous shells of organisms form layers of growing thickness by layer-upon-layer accumulation of particles.

Size classification

A simplified particle-size classification in wide use for clastic sedimentary rocks is shown in Table V. A somewhat different size classification commonly is used for pyroclastic rocks, as shown in Table VI.

TABLE V

Particle-size classification for clasts in sedimentary rocks

	"Average" diameter (mm)	*Typical rock name*
Coarse-grained	> 2	conglomerate
Medium-grained	2–0.0625 (2–$\frac{1}{16}$ mm)	sandstone
Fine-grained	< 0.0625 (< $\frac{1}{16}$ mm)	shale, siltstone
silt-size	0.0625–0.004	
clay size	⩽ 0.004	

TABLE VI

Particle-size classification for pyroclastic rocks

Particles	*Average size* (mm)	*Name of aggregate*
Blocks, bombs	> 32	volcanic breccia, agglomerate
Lapilli	32–4	lapilli tuff
Coarse ash	4–¼	coarse tuff
Fine ash	< ¼	fine tuff

Shapes of clasts

Shapes of clastic particles may be identified with sufficient accuracy for most purposes by giving consideration to relative dimensions such as length, breadth, and width and by noting the extent to which edges and corners of the particles are rounded by solution or abrasion. A satisfactory terminology for shapes based on relative dimensions is exactly the same as the terminology for habits in minerals (Fig. 4). By observing the roundness (or lack of it) of corners or edges on fragments of any habit, use can be made of the expressions *angular, subangular, subrounded, rounded,* and *well-rounded.*

Particle orientation

Clasts with nonequidimensional shapes tend to arrange themselves in definite patterns during sedimentation and contribute to a great extent to the physical properties of rocks. For example, deposition of platy or micaceous minerals such as the clay minerals or micas parallel or subparallel to stratification renders rocks easily susceptible to splitting or cleaving in the direction of the stratification.

Particle orientation ranges from random to highly preferred, that is, particles with shapes other than equant tend to develop a strong element of parallelism or subparallelism during deposition.

Stratification (bedding)

Clastic particles deposited from suspensions or solutions build up layers. Changes in the rate of deposition and changes in the sizes and mineralogy of the deposited materials result in *stratification* or *bedding*. When materials are dumped at a rapid rate, slumping ensues and planar stratification becomes less evident as internal fabrics become very complex.

Bedding in stratiform deposits can be characterized as *regular*, when contacts between adjacent layers are flat or relatively smooth, or *irregular*, when stratiform relationships are complex or not even apparent, especially when slumping in thick accumulations has altered simple geometrical relationships.

Stratification can be described by several terms such as laminar (very thin beds), thin-bedded, thick-bedded, or actual thicknesses may be specified.

Matrix and cement

The process whereby clastic accumulations of loose particles are converted into rocks is called *lithification*, and follows at any time after deposition or almost at the same time. Clasts are deposited initially as loose aggregates with more or less interstitial pore space. Finer-grained materials may be deposited simultaneously to form a *matrix* for larger clasts, or simultaneously with deposition or at a later time chemically precipitated substance may occupy the pore spaces as *cement*. Further, a matrix by subtle changes involving changing crystallinity or compaction by loss of water may be converted to a cement.

Cement may be formed from existing materials by solution and redeposition or may be introduced from an outside source by circulating solutions. In any event competency in lithified clastic accumulations depends to an important degree on the nature of the matrix and/or cement. For example, a siliceous sandstone with a silica cement is a much more competent rock than a sandstone held together by a dehydrated clay matrix.

A special situation prevails in many accumulations of pyroclastic rocks which were explosively erupted as hot particles of volcanic glass, sometimes accompanied by hot, burning gases. Clasts in these accumulations, still hot and semiplastic, tend to flow under the weight of overlying layers, and become welded, producing intricate, interlocking patterns of articulation of the clasts. Such rocks, whether welded or not, are classified as *ignimbrites*.

IGNEOUS ROCKS

Igneous rocks form by cooling and solidification of hot, completely to partly molten more or less viscous solutions of silicates. Depending on intitial chemical composition and the rate of cooling, glassy, very fine-grained, fine-grained, medium-grained, coarse-grained, or very coarse-grained fabrics are produced. In some igneous rocks, called *porphyries* there is a notable difference in the sizes of the mineral grains in a rock.

The most abundant minerals in igneous rocks are the feldspars. Next in importance is quartz, and of secondary importance are the feldspathoids. In addition to these minerals most igneous rocks contain other minerals such as the micas, amphiboles, and pyroxenes, and a large variety of minerals which are generally present in amounts less than two or three per cent by volume.

Essential minerals in igneous rocks are those whose presence is necessary in classifying the rock. *Varietal accessory* minerals are minerals identifying varieties. For example, a *granite,* by definition, contains abundant feldspar and quartz, but if biotite also is abundant, the granite may be classified as a *biotite granite.*

A complete classification of igneous rocks based on chemical composition or detailed mineralogy determined by petrographic examination of thin sections, or by other means, is an elaborate, quantitative characterization of rocks. For engineering purposes, a simplified classification based on megascopically visible properties is quite sufficient.

The key in Table VII is based on fabric and mineral composition as observed in hand specimen or field exposures. Igneous rocks with notably larger crystals embedded in a finer-grained or glassy matrix are called *porphyries;* the larger crystals are *phenocrysts*; and the finer-grained matrix is *groundmass.* Porphyries sometimes are given rock names with the suffix *-phyre.*

SOME SHAPES OF IGNEOUS BODIES

Igneous rocks form by the crystallization of masses of molten rock derived from the deep interior of the earth or by partial to complete fusion of pre-existing rocks in the earth's crust. A distinction may be made between *intrusive bodies,* which are injected into the rocks beneath the earth's surface, and *extrusive bodies,* which build up accumulations on the earth's surface. A further distinction may be made between *discordant bodies,* which intersect intruded masses of rock, and *concordant intrusives,* which are in structural harmony with the intruded rocks.

In general, small or thin bodies of igneous rocks which have congealed quickly are glassy, fine-grained, or porphyritic. Larger bodies, which have cooled slowly tend to be coarser grained. Several examples of large and small igneous bodies are illustrated in Fig. 10–13.

TABLE VII

Key for megascopic classification of igneous rocks

I. Rock is glassy and
 A. Contains few or no phenocrysts:
  obsidian (vitreous luster)
  pitchstone (dull or waxy luster)
 B. Contains conspicuous or abundant phenocrysts:
  vitrophyre, either *pitchstone vitrophyre* or *obsidian vitrophyre*

II. At least 50% of rock is very fine-grained and outlines of crystals are below threshold of visibility.
 A. Rock is light-colored and entirely very fine-grained:
  felsite
 B. Rock is dark-colored and entirely very fine-grained:
  basalt or *trap*
 C. Rock is light-colored and contains abundant phenocrysts:
  1. Quartz and feldspar can be recognized in phenocrysts:
   rhyolite porphyry
  2. Feldspar, but no quartz can be recognized in phenocrysts:
   trachyte porphyry
 D. Rock is dark-colored and contains abundant phenocrysts of tabular feldspar:
  basalt porphyry

III. At least 50% of rock is coarse-grained enough to allow visual examination of individual grains.
 A. Rock is light-colored and approximately equigranular:
  1. Quartz and feldspar are abundant:
   granite
  2. Feldspar is abundant but no quartz is observed:
   syenite
 B. Rock is light-colored and porphyritic:
  1. Quartz and feldspar are abundant:
   granite porphyry
  2. Feldspar is abundant but quartz is absent:
   syenite porphyry

IV. Rock is medium-to-coarse-grained and contains abundant dark-colored accessory minerals. Feldspar is generally tabular and shows twin striations on cleavage surfaces:
 gabbro
 (*anorthosite* is a variety of gabbro consisting almost entirely of calcic plagioclase)

V. Rock consists of very large (1 inch or more) crystals, commonly quartz and feldspar:
 pegmatite
 (sugary-grained quartz-feldspar aggregates commonly associated with pegmatite are called *aplite*)

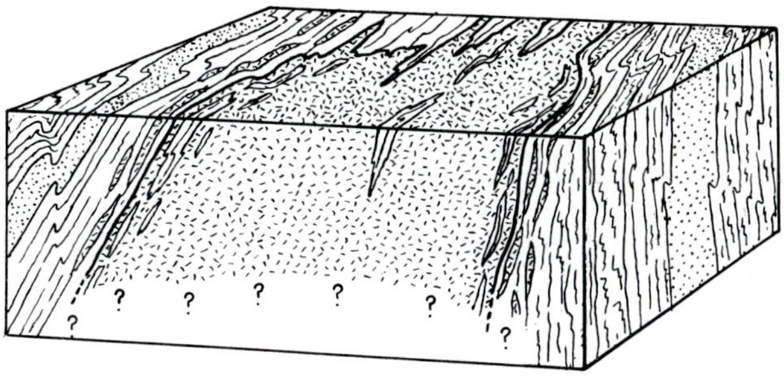

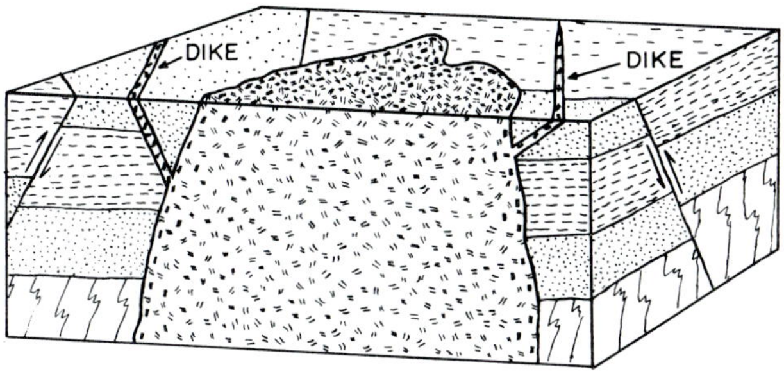

Fig. 10. Some large intrusive igneous bodies.
A. Batholith of granite intruding metamorphic rocks. Area of surface exposure exceeds 40 sq. miles (100 km^2). Projection of batholith downward is uncertain. Many batholiths have distinct floors.
B. Stock intruding faulted metamorphic and sedimentary rocks. Apophysal dikes intersect country rock and follow faults. Area of surface exposure of stock is less than 40 sq. miles (100 km^2).

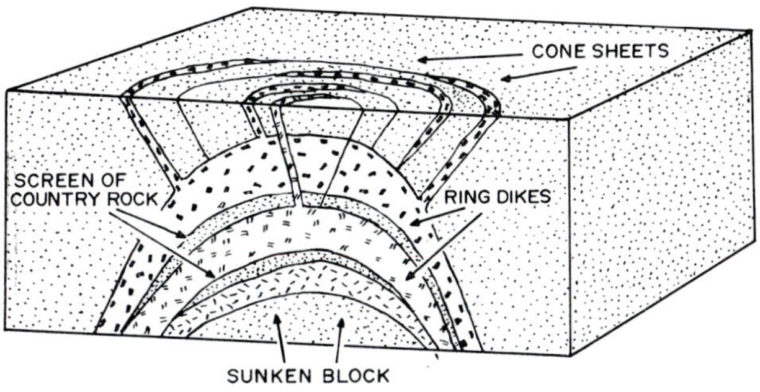

Fig. 11. Ring dikes and cone sheets formed as a consequence of subsidence of a dome-shaped block into a magma reservoir. Many stocks probably have formed by the mechanism implied in the illustration.

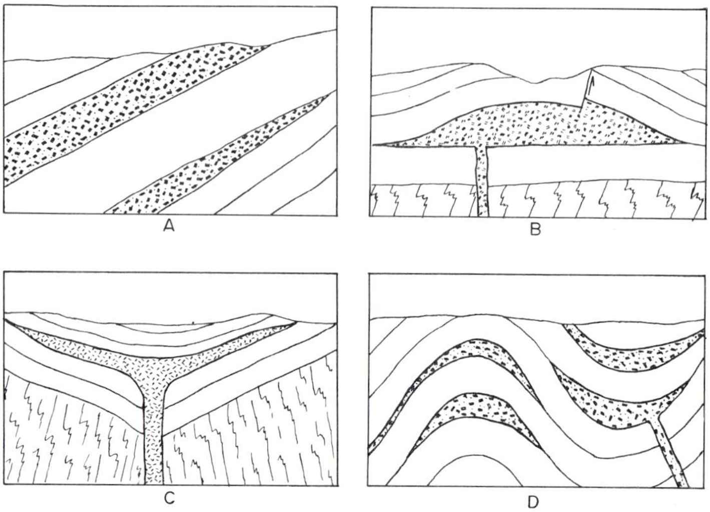

Fig. 12. Concordant igneous intrusions. A. Sills injected along bedding planes of tilted sedimentary rocks. B. Laccolith injected into sedimentary rocks and causing doming of overlying layers. A fault has formed as a consequence of intrusive pressures. C. A lopolith, a dish-shaped intrusive into layered sedimentary rocks. D. Phacoliths formed by intrusion of magma into crest of anticline and trough of a syncline.

Fig. 13. Extrusive igneous bodies. A fissure eruption fed by a subterranean dike spreads out over the land surface, and volcanic cones are built up by lava flows and pyroclastic rocks moving toward the earth's surface along volcanic necks of approximately circular cross-section.

SEDIMENTARY ROCKS

Sedimentary rocks can be classified into two broad categories: (1) dominantly clastic rocks held together with a compacted matrix or a chemical cement such as silica, carbonates, or iron oxides, and (2) dominantly crystalline rocks such as rock salt, gypsum, anhydrite, and certain crystalline limestones and dolostones, which were constituted into crystalline aggregates at the time of or slightly later than their accumulation in strata, or which recrystallized at a considerably later time owing to deep burial and/or by activity of ground water.

Table VIII is a key for megascopic identification of common sedimentary rocks. Rocks of mixed fabric or composition can be further classified by using appropriate modifiers, for example, *calcareous shale, sandy shale, clayey sandstone,* etc.

TABLE VIII

Key for megascopic identification of common sedimentary rocks

I. Rock is dominantly an accumulation of clastic silicate particles or rocks bonded by a matrix and/or a cement.

A. Average size of particles is clay- or silt-size. Clay-size particles are below the threshold of visibility in hand-specimen and even under the microscope. Silt-size particles are below or just at the threshold of visibility with a hand lens, but are so small that identification is not generally possible.

claystone
mudstone
shale (bedded and fissile)
siltstone

B. Average size of clastic particles is sand-size (about 0.06–2 mm):

sandstone

1. Particles are angular and consist of a variety of minerals, especially feldspar and quartz. Matrix contains clay minerals.

graywacke sandstone

2. Particles are angular and consist mainly of feldspar and quartz. Clay matrix is absent.

arkosic sandstone

3. Particles are rounded and consist mainly of quartz. Cement is generally silica or carbonate.

quartzose sandstone

C. Average size of particles is greater than 2.0 mm or conspicuously larger rock or mineral fragments are present in a finer-grained aggregate:

1. Particles are angular:

conglomerate breccia

2. Particles are more or less rounded:

conglomerate

TABLE VIII (continued)

II. Rock is dominantly an accumulation of clastic particles of carbonate minerals:
- 1. Very fine-grained:
  - *micritic limestone* (consists of tiny calcite particles and generally requires microscopic examination before appropriate identification)
- 2. Particles are of sand size (0.06–2 mm):
  - *calcarenite* (calcite)
  - *doloarenite* (dolomite)
- 3. Average size of particles is greater than 2.0 mm or conspicuously larger fragments are present in a finer-grained aggregate:
  - *limestone breccia*
  - *dolomite breccia*
  - *limestone conglomerate*
  - *dolomite conglomerate*

III. Rock consists dominantly of anhedral to euhedral crystals and fabric is nonclastic. Use grain-size classification Table IV.
- *crystalline limestone* (calcite)
- *crystalline dolostone* (dolomite)
- *gypsum*
- *anhydrite*
- *rock salt*

IV. Rock is pyroclastic, having formed by accumulation of particles of explosive volcanic origin, crystals and/or glass.
- Particles greater than 32 mm in diameter:
  - *volcanic breccia; volcanic agglomerate*
- Particles 32–4 mm in diameter:
  - *Lapilli tuff*
- Particles 4–$^1/_4$ mm in diameter:
  - *Coarse tuff*
- Particles less than $^1/_4$ mm in diameter:
  - *tuff*; if mainly glassy: *vitric tuff*

METAMORPHIC ROCKS

Metamorphic rocks have resulted from partial to complete recrystallization at high temperatures and/or pressures. Depending on the intensity of metamorphism, it is possible to distinguish low-grade, medium-grade, and high-grade mineral assemblages. Under many conditions metamorphic rocks are notably deformed during recrystallization and assume characteristic, sometimes very complex fabrics. Layering, often called *foliation* or *schistosity,* is conspicuously displayed in hand specimen or field exposures.

The most useful megascopic classification is based on visible fabric and mineralogy. An outline classification is given in Table IX.

TABLE IX

Outline megascopic classification of metamorphic rocks

I. Mineral components are randomly oriented in hand specimen, whether platy, elongate, or equant. Foliation not apparent or very poorly developed.

quartzite (mainly quartz)
marble (mainly calcite)
dolomitic marble (mainly dolomite)
hornfels (fine-to-coarse-grained, mainly silicate minerals)

II. Rock consists mainly of platy, micaceous, prismatic, or acicular minerals in subparallel to parallel arrangement.

1. Very fine-grained. Grains below limit of visibility in hand specimen. Rock generally splits easily into thin plates.
   slate
2. Fine-grained. Some grains can be seen with hand lens. Splitting tendency often pronounced.
   phyllite
3. Particles are large enough to be observed in hand specimen.
   schist, especially *biotite schist, muscovite schist,* and *hornblende schist*

III. Rock contains abundant quartz and/or feldspar and layering or linear arrangement of component minerals is not especially pronounced as in schist. In some rocks and on a small scale layers rich in *quartz* and/or *feldspar* alternate with layers rich in platy, micaceous, prismatic, or acicular minerals.

gneiss, especially *biotite gneiss, muscovite gneiss,* and *hornblende gneiss*

IV. Rock appears to be a complex intermixture of metamorphic rocks and granular igneous rocks.

migmatite

REFERENCES

Heinrich, E.W., 1965. *Microscopic Identification of Minerals.* McGraw-Hill, New York, N.Y., 414 pp.

Moorhouse, W.W., 1959. *The study of Rocks in Thin Section.* Harper and Row, New York, N.Y., 514 pp.

Wahlstrom, E.E., 1969. *Optical Crystallography.* Wiley, New York, N.Y., 4th ed., 489 pp.

Chapter 3

ROCK ALTERATION

INTRODUCTION

Rocks and the minerals in them are subject to a variety of physical and chemical changes during their lifetimes. Each rock is a consequence of a combination of physicochemical forces acting together in a certain manner at some time in the geologic past. The fabric and mineralogical composition of a particular rock reflect an effort on the part of the mineral aggregate comprising the rock to attain a condition of at least approximate thermodynamic equilibrium with the geological environment of formation of the rock. Thus, it becomes possible to specify limiting physicochemical conditions for the formation of the various kinds of rocks. For example, igneous rocks consolidate by cooling through a range of high temperatures at various depths beneath the earth's surface or on the earth's surface. Sedimentary rocks are deposited at low temperatures beneath the atmosphere or under water and have physical and chemical characteristics that are diagnostic of the environment of deposition.

Some minerals, for example quartz and calcite, are stable through a wide range of geological environments. Others, such as olivine and calcium plagioclase, are not. Moreover, where a succession of minerals crystallizes through a range of temperatures as in metamorphism or during cooling of a magma to form an igneous rock, minerals that are stable only in narrow ranges of high temperature, obviously can continue to persist only in a nonequilibrium state. From the above argument it is seen that igneous and metamorphic rocks exposed at the earth's surface or buried rocks thrust into new environments generally are thermodynamically unstable and possess the capability for more or less extensive changes in mineralogy and fabric.

The most efficient agent promoting changes in the direction of attainment of equilibrium is water, and most of the notable alteration changes in rocks occur only in the presence of water. Included are the extensive alterations producing weathered rocks at or near the earth's surface, and a great variety of alterations that occur at shallow to great depths in the presence of warm to hot water. Temperature plays an important part in that chemical reactions induced by aqueous solutions progress much more rapidly at high temperatures than at lower temperatures.

From an engineering point of view some of the most serious problems in rock excavation, especially underground, are directly the consequence of alteration of previously competent rocks so as to reduce their competency, at times to such an extent that the altered material behaves plastically under pressure or swells and disintegrates to a mass of slurry-like consistency in the presence of ground water or atmospheric moisture.

Many alteration processes reduce the average grain-size of rocks. An outstanding example is weathering alteration that converts feldspar crystals to clay aggregates of micron size. Alterations do not always reduce the competency of a rock or the average grain-size. Introduction of silica into a rock with an increase in grain-size under certain circumstances may improve the hardness and cohesiveness of an initially incompetent rock considerably.

The techniques for study of alteration in rocks are the same as those for examination of unaltered rocks as described in the previous chapter.

WEATHERING

Weathering is a process causing changes in rocks at and near the earth's surface by interaction between the rocks and chemically active components of the earth's atmosphere, principally water, carbon dioxide, and oxygen. The effects of weathering extend below the earth's surface by ground-water movement through or around grains or along joints and fissures. Unless the products of weathering are removed by erosion, weathering tends with time to produce a mineral aggregate in equilibrium with the surface weathering agents, and the depth of weathering is limited by an armoring action of the weathered layer.

The depth of weathering is largely controlled by topography and the availability of channels of flow for surface waters. In some regions where climatic, topographic, and structural conditions are appropriate, weathering caused by circulation of ground waters of surface origin extends hundreds to thousands of feet underground and causes notable changes in originally fresh rocks. Weathered rocks from an engineering point of view generally are notably less competent than original rocks, and the changes in rock properties due to weathering may have considerable bearing on procedures that must be followed in underground excavations, particularly in relatively shallow excavations or in tunnel portal areas.

Within a given area on the earth's surface and at a given geologic time, the particular effects of weathering must be assessed in terms of at least five variables: (1) climate, (2) biological activity, (3) topography, as it effects underground movement of water, or contributes to erosional removal of weathered material, (4) parent material, and (5) time.

The geologist concerned with the possibility of the existence of weathered rocks at depth is confronted with a complex variety of considerations not the least of which is the knowledge that weathered zones have developed throughout most of geologic history and, in many instances in environments considerably different from present environments. Many deeply weathered zones developed in the geologic past were stripped away by erosion, but, in many locations, beneath major surfaces of unconformity representing intervals of long exposure of rocks at the earth's surface, weathered rocks are preserved under overlying younger rocks with little or no change in original characteristics. Detection or prediction of such zones requires skilled interpretation of geologic history. Buried weathered zones may exist in igneous or metamorphic rocks beneath younger successions of sedimentary rocks or may exist below unconformities within a sedimentary succession.

The end product of weathering generally is an oxidized mineral residuum and aqueous pore solutions containing elements derived from the weathering process. The rate and intensity of weathering in a given rock are dependent on temperature and the availability and freedom of movement of surface water within the rock. Rate and intensity also depend on the mineralogical composition and fabric of the rock.

Mechanical weathering

The mechanical breakdown of rocks at the earth's surface, largely because of expansion and contraction with short term and long term temperature changes, and abetted by frost action and expansion caused by roots of organisms, is called *mechanical weathering.* It is effective in promoting chemical weathering in the upper few feet or tens of feet of rocks exposed at the earth's surface.

Weathering related to mineral composition

Minerals that are to the greatest extent in thermodynamic disequilibrium with the weathering environment are the first to break down chemically. Among the common minerals in igneous and metamorphic rocks or in particles in clastic sedimentary rocks several groups with differing resistances to weathering may be identified (Table X). It follows that rocks rich in minerals with low resistance to weathering decompose more readily than in rocks in which these minerals are absent or present in minor amounts.

In general the minerals most susceptible to weathering contain abundant magnesium, calcium, and iron. The iron generally remains in a rock after oxidation and hydration, whereas the magnesium and calcium and, later the alkalies, are removed by flushing or by diffusion through solutions along concentration gradients or are fixed as carbonates. The least mobile elements in weathered rocks are aluminum, silicon (silica), and iron, and the commonest kind of chemical breakdown during weathering yields clay minerals or closely related fine-grained micaceous minerals plus hydrated oxides of aluminum and iron. In environments of extreme leaching, particularly in tropical regions even silica is leached, and only iron oxides and aluminum hydrates remain to form so-called *laterites.*

TABLE X

Minerals of igneous and metamorphic rocks: resistance to chemical weathering

1. Low resistance:	
olivine	staurolite
pyroxene	iron-rich garnet
hornblende	melilite
calcic plagioclase	epidote
biotite mica	feldspathoids
2. Intermediate resistance:	
intermediate plagioclase	sillimanite
sodic plagioclase	andalusite
orthoclase feldspar	kyanite
microcline feldspar	calcium aluminum garnet
muscovite mica	iron-poor garnet
	clinozoisite
3. High resistance:	
quartz	hematite
magnetite	tourmaline

The first manifestations of weathering in feldspar-rich rocks are in breakdown of iron–magnesium silicates and conversion of feldspars into clay minerals (Fig. 14) to produce a dull-looking, crumbly, iron-stained loose material often given the name *saprolite*. In such rocks original fabrics are still recognizable, and volume changes by extensive leaching are not notable. Saprolite is widespread beneath the upper horizons that develop during formation of soils.

Weathering related to ground-water movement

Removal of dissolved products from weathered rocks is accomplished by diffusion through pore water or by actual transport by water moving through the rock under a head or by capillarity. Movement by diffusion or by capillary transport involves relatively small distances. Depending on permeability and the presence or absence of a potential for active circulation of ground-water distinctions may be made between stagnant environments and environments that are effectively leached by moving ground water.

In stagnant environments, where diffusional transport is the only active mechanism for leaching, the products of weathering tend to retain calcium, magnesium, and the alkalies, and the abundant clay minerals commonly are montmorillonite, illite, chlorite, vermiculite, and mixed layer clays. Continued removal of the alkalies and alkaline earths by actively circulating ground water leaves clay mineral residues rich in kaolinite, and in extreme situations, residues of aluminum hydroxides (gibbsite) and iron oxides.

Fig. 14. Pervasively altered granite. Alteration of feldspars to clay minerals has produced an aggregate easily crushed in hands. Rounded masses are relatively unweathered residuals showing spheroidal weathering. Small dikes are aplite.

It is apparent that weathering at depths of hundreds or thousands of feet requires active movement of ground water of surface derivation (containing oxygen and carbon dioxide and perhaps organic chemicals) presumably along fractures or other kinds of permeable zones, perhaps sedimentary aquifiers. With these conditions specified, deep weathering might be expected in regions of considerable topographic relief and wet, temperate to hot climate.

Engineering properties of weathered rocks

From the foregoing discussion it is seen that weathered rocks encountered underground may display a variety of properties ranging from behavior as loose aggregates of larger particles to behavior as plastic, very fine-grained or clayey rocks easily converted into dispersed aqueous suspensions or increasing in volume by absorption or bodily incorporation by expanding clay minerals of available water.

LOW-TEMPERATURE ALTERATIONS AT MODERATE DEPTHS

Because of the geothermal gradient waters circulating at some distance below the earth's surface can be expected to have somewhat higher temperatures than near surface waters. This section is concerned with alterations brought about in rocks at a few thousands of feet in depth, at temperatures of the order of 50–100°C and under conditions where hydrostatic head or rock-load pressures exceed those normally associated with weathering processes at and near the earth's surface. Ground water at such depths is capable of circulating through available channelways, if topographic conditions on a large or small scale produce a hydraulic gradient. In addition to circulating water there may be relatively stagnant pore water which may promote local chemical reactions by enabling diffusion of dissolved chemicals on a relatively small scale. The changes in rocks at moderate depths, especially the changes in sedimentary rocks, are frequently referred to as *diagenetic changes.*

Clayey aggregates tend to lithify by compaction and loss of pore water. Clastic accumulations become more competent by compaction of fine-grained matrix and by precipitation of cement. In addition, especially in limestones, anhydrite, and salt deposits, recrystallization produces coarser-grained fabrics, or moving solutions introduce new substances such as silica into the rock. In very fine-grained rocks pore water may induce local transfer of dissolved chemicals to cause minor to major changes in mineralogy.

The changes in mineralogy and fabric under the conditions specified above tend to increase cohesiveness, strength, and average competency of the affected rocks.

HYDROTHERMAL ALTERATION

Within a temperature range of about 100–500°C and at depths ranging from shallow to great hot aqueous solutions derived from a source such as a cooling igneous body exert profound influences on the mineralogy and fabric of invaded rocks and create some of the most hazardous conditions encountered in underground excavations.

Much is known concerning hydrothermal alteration of rocks, because such alteration is a common accompaniment of deposits of economically important metallic minerals in ore deposits. However, in many areas hydrothermal alteration has changed large volumes of rock without formation of metalliferous deposits.

The most spectacular examples of hydrothermal alteration are found in regions of intrusive and extrusive igneous activity, and there is a close space and time relationship between the igneous activity and the alteration. Theory and observation indicate that concentrations of hydrothermal solutions accumulate in the late stages of an igneous cycle so that alterations spread out in the rocks adjacent to congealed igneous bodies but, also, may change mineralogies and fabrics in the igneous bodies themselves.

Extensive hydrothermal alteration generally is controlled by major or minor channelways of circulation localized by faults or joints. On a smaller scale, alteration spreads out in rocks by movement of solutions along grain boundaries of minerals in aggregates or by solid diffusion through crystals.

A common phenomenon is volume-for-volume replacement of existing minerals by new minerals growing from elements introduced into a rock from an outside source; a process described as *metasomatism*.

Limestone is among the rocks most susceptible to alteration by hydrothermal alteration and shale, mudstone, and sandstone are among the rocks least susceptible to such alteration. In general, rocks containing abundant alkaline earths and iron are more likely to be altered by hydrothermal solutions than rocks containing abundant silica and alkalies.

Several kinds of hydrothermal alteration have been identified. The important kinds and the changes they effect are described in subsequent paragraphs. Although the various kinds of alteration are described separately, two or more kinds commonly are observed in a particular area, and frequently various kinds of alteration show distinct zonal arrangements with respect to major or minor avenues of underground circulation of the hydrothermal solutions.

Silicification

Silicification results from the addition of silica, generally in the form of quartz or chalcedony into rocks. Especially noteworthy is the extensive replacement of calcareous rock by silica and the conversion of all kinds of rock to nearly pure quartz adjacent to fissures. Silicification tends to improve the competency of rocks by providing an interlocking aggregate of hard, resistant quartz grains. Some silicified rocks assume the aspect of highly indurated sandstones or metemorphic quartzites.

Sericitization

Sericite is very fine-grained muscovite mica or its sodic counterpart, paragonite. Closely related to sericite is *illite,* a clay mineral, and much so-called sericite is illite, a substance containing more water and smaller amounts of potassium than true sericite.

Sericitization is a highly selective process and generally results in replacement of feldspars, first intermediate and calcic plagioclase feldspars and then alkali feldspars. Quartz is resistant to sericitization, and it is not uncommon for silicification and sericitization to operate simultaneously.

Rocks which have been pervasively sericitized are dull or waxy appearing and tend to have low strengths. Under directed pressures sericitic rocks behave in almost the same manner as fine-grained clayey rocks.

Argillization

Argillization is a process whereby rocks are converted by hydrothermal alteration to clay mineral aggregates. Chiefly affected are all of the common rock minerals except quartz. The minerals produced by argillization are the same as those resulting from weathering and the particular minerals that are formed depend on the extent of leaching of alkalies and alkaline earths from the altered rocks. With imperfect leaching chlorite and montmorillonite replace preexisting silicate minerals and with total extraction of soluble cations kaolinite is left as a residue.

Argillization produces one of the most treacherous of all conditions encountered in underground excavation by reducing previously competent rocks to practically incoherent, squeezing, and swelling aggregates of clay minerals when exposed to atmospheric or ground water (Fig. 15 and 16).

Pyritization

Introduction of pyrite into rocks by hydrothermal solutions produces massive deposits or scattered crystals, sometimes forming at great distances from the source. Pyritization commonly is accompanied by other kinds of alteration and the presence of pyrite crystals provides a warning that the associated rocks have been penetrated by solutions which may have caused chemical breakdown and replacement by silica, sericite, or clay minerals.

Propylitization

Propylite is an altered rock, commonly of volcanic origin, which has been notably affected by hydrothermal solutions. The rock usually has a greenish cast and

Fig. 15. Thoroughly argillized schist and gneiss. Hydration has caused extensive swelling pressures. Pilot bore, Straight Creek Tunnel, Colorado. (Photo by Colorado Division of Highways.)

upon microscopic examination reveals the abundant presence of chlorite, epidote, and to a variable extent clay minerals and carbonates. Propylitization reduces the competency of original rocks, but not generally to the same extent as pervasive argillization.

Chloritization

Hydrothermal solutions commonly convert aluminous ferromagnesian minerals such as pyroxene, hornblende, and biotite to greenish aggregates of fine-grained chlorite. Chloritization commonly is accompanied by other kinds of alteration, especially argillic alteration.

Fig. 16. Mud squeezing through lagging in section of fractured, argillized granite. Vasquez Tunnel, Colorado. (Photo by Denver Board of Water Commissioners.)

Dolomitization

Rocks containing abundant calcite (limestones) are highly susceptible to replacement by dolomite and are described as dolomitized. Dolomitization is a low-temperature sedimentary process as well as a hydrothermal process. In many metalliferous districts dolomitization of limestones is widespread and extends for thousands of feet or even miles along limestone layers. Frequently, dolomitization produces small crystals with characteristic rhomb-shaped outlines.

Many dolomitized limestones are rendered vuggy and porous by dolomitization by solutional activity of penetrating solutions. In rare instances dolomitization produces loose aggregates of sand-size crystals that possess no cement and flow freely into underground openings.

Saussuritization

Saussurite is the name given to a fine-grained, commonly closely sheared and fractured rock, which has a complex composition resulting from partial to complete alteration of original minerals, especially calcic plagioclase. Abundant minerals resulting from chemical breakdown of original minerals are epidote or clinozoisite, sericite, carbonates, and sodic plagioclase. Extensive saussuritic alteration usually produces a compact rock with the appearance and the physical characteristics of a fairly well-lithified mudstone.

DEUTERIC ALTERATION

Igneous rocks form from magmas, completely to partly molten solutions of silicates. Water is present in the magma in amounts of a few per cent by weight. Crystallization of magmas generally is initiated by formation of anhydrous silicates and, by subtraction of these silicates, the magma is enriched in water. In late stages of crystallization some of the water is incorporated as hydroxyl in minerals such as hornblende and biotite, but when crystallization is completed there is usually a water-rich residual solution which permeates the rock or escapes into surrounding country rocks. The temperatures of these solutions is high, of the order of 500–700°C.

The residual solutions have the capacity to convert preexisting minerals into new ones by so-called *deuteric alterations.* Widespread products of deuteric alteration are hornblende formed at the expense of pyroxene, chlorite, epidote, zoisite, and albite. In deuterically altered igneous rocks rich in magnesium serpentine and talc develop, and in rocks deficient in silica zeolites tend to form. In some rocks quartz replaces feldspars in close, complex intergrowths.

Deuteric alteration rarely affects the physical properties of igneous rocks, and in some instances where the solutions have deposited minerals in joints or cracks resulting from cooling, the properties of the rock may be improved.

PNEUMATOLYTIC ALTERATION

Volcanic activity at the earth's surface resulting in eruption of lava flows and pyroclastic materials is accompanied by generation of large volumes of hot gas, principally steam, but also containing a variety of other substances including combustible compounds, carbon dioxide, acids, and the halogens chlorine and fluorine. The alterations that result from interaction between the gases and country rock or ejected materials are described as *pneumatolytic alterations.* Familiar examples of pneumatolytic alterations are observed around fumaroles in volcanic regions.

With a decrease in volcanic activity and as a consequence of cooling gaseous emanations condense and gradually give way to hot spring activity.

Hot gases and aqueous solutions penetrating explosive ejectamenta near volcanoes, especially porous, unconsolidated pyroclastic accumulations often produce pervasive alterations especially argillic alteration. Altered pyroclastic accumulations even after burial and compaction tend to disintegrate and swell on contact with water or water vapor and present a very difficult challenge in tunneling operations.

MISCELLANEOUS TYPES OF ALTERATION

In addition to the foregoing types of alteration, there are additional types which have been noted in rocks. Generally, the alterations mentioned in this section do not notably decrease the grain size or competency of the altered rocks, and in some kinds the quality of a rock as regards its characteristics during excavation is notably improved. Included among the miscellaneous types of alteration are the following:

(1) *Feldspathization* whereby feldspar is generated in a rock by introduction from an outside source or by mobilization of existing materials.

(2) *Zeolitization*, which results in alteration of preexisting minerals at low to moderately high temperatures or results from crystallization of introduced materials.

(3) *Carbonatization,* which is a consequence of introduction of calcite into a rock.

(4) *Chertification,* in which very fine-grained silica is introduced into a rock.

(5) *Tourmalinization,* resulting in formation of tourmaline needles by replacement of preexisting minerals, generally at high temperatures.

(6) *Epidotization,* resulting in the crystallization of epidote.

(7) *Garnetization,* whereby garnets replace preexisting minerals.

It should be understood that, whenever a new mineral is formed in a rock, either by introduction from an outside source or crystallization from components locally available in a rock, the generation of the new mineral may be regarded as a kind of rock alteration.

Chapter 4

ELEMENTARY ROCK MECHANICS

INTRODUCTION

Since man first started to excavate underground openings he has been forced to give consideration to the behavior of rocks under loads of overlying rocks. Much practical experience was gained through the centuries in dealing with adverse underground conditions, especially as subsurface mining of economic mineral deposits began to flourish. Geologists during the past hundreds of years have been concerned with the stresses in the earth's crust that have deformed and ruptured rocks, and, by regional world-wide studies and by detailed local studies have identified the space and time relationships of large and small rock dislocations. Such studies inevitably have led to numerous conjectures concerning the distribution and intensities of the forces that caused the dislocations, and many broad concepts of rock mechanics based on empirical observations have become the geologist's stock-in-trade.

Traditionally engineers have beeen concerned with physical properties of artificial and natural materials, especially their behavior under stress. During the first part of this century engineers turned their careful attention to measurement of physical properties of rocks and established quantitative criteria for failure of plastic and elastic materials. Accompanying the development of quantitative measurements was the establishment of an increasingly elegant theoretical approach.

In the early days data accumulated in the laboratory concerning rock behavior were of relatively slight concern to the geologist or the mining engineer, because the rock bodies that were of interest to him were tremendously more complex than the simple blocks and cylinders subjected to squeezing, tension, and torque in the laboratory.

In the 1960's papers on rock mechanics began to appear in profusion, with contributors representing several areas of specialization, including geology, mining, petroleum technology, and civil engineering. Rock mechanics clearly had become identified as a field of very wide interest, and not the domain of any single group of specialists.

Rock mechanics is concerned with all aspects of the behavior of rock under stress and is basic to the planning, design, and construction not only of underground excavations, but also of excavations in rock at the earth's surface. Soil mechanics, a related field, is more concerned with the behavior of generally unconsolidated natural materials, but no sharp distinction can be drawn between the behavior of some soils and the behavior of certain chemically altered or mechanically disintegrated rocks of low strength encountered in underground excavations.

GLOSSARY OF DEFINITIONS

The language of rock mechanics has grown rapidly through the years, and the following basic definitions are intended to acquaint the reader with the meanings of many terms currently used in elaborations of theory and technology of rock mechanics. Included are some terms that are not used in text discussions.

Body force. A force acting throughout a body of material produced without contact of the body with other bodies. Examples are magnetic, gravitational, and inertial forces.

Competent rock. A rock which is capable of sustaining an underground opening or a steep slope at the earth's surface without artificial support.

Compressive strength. A measure of the strength of a material under compression.

Coulomb failure. Failure of material when the maximum shear stress at a point in a material reaches a critical value called the *shear strength.*

Creep. Time-dependent deformation or strain.

Ductile behavior. As in metals such as copper which do not rupture at a definite stress, but, instead, maintain a cohesive strength over a very large stress range.

Elastic solid. A solid in which the stress σ (sigma) is directly proportional to the strain ϵ (epsilon) as expressed by the relationship $\sigma/\epsilon = E$, where E is Young's modulus, a constant for each elastic material.

Firmo-viscous material. A material in which with time the strain relaxes under zero stress.

Fracture. A surface of discontinuity formed by rock failure under stress.

Hookean solid. A solid in which the strain (ϵ) is directly proportional to the stress (σ) as in the relationship $\sigma/\epsilon = E$, where E is Young's modulus.

Hooke's law. A statement that strain is directly proportional to stress.

Incompetent rock. Rock incapable of standing in underground openings or steep slopes at the surface without support.

Isobars. Curves of constant principal stress.

Isopachs. Curves of constant mean stress.

Kelvin material. Firmo-viscous material, which see.

Maxwell material. Viscoelastic material with combined elements of elastic and viscous (fluid) materials.

Modulus of elasticity. Young's modulus, which see.

Mohr's failure. Failure of a material by fracture or permanent deformation when the shear stress on a real or potential fracture of slip plane has increased to a value which depends on the stress acting normal to the plane or when the maximum principal tensile stress has reached a certain limiting value.

Newtonian material. A fluid which can not withstand a shearing stress without deforming permanently.

Plastic behavior. The behavior of plastic material, which see.

Plastic material. Material which deforms permanently when an applied stress has reached a characteristic critical value.

Poisson's ratio. A ratio expressed by the symbol γ (gamma) such that $\gamma = \epsilon\, E/\sigma$ where ϵ is strain, E is Young's modulus, and σ is stress. Poisson's ratio for many materials is about 0.25.

Rupture. Failure of a material with development of fractures or shears.

Shear. Displacement of adjacent elements along a flat or curved surface.

Shear strength. A measure of the resistance of a material to shear stress.

Shear stress. A stress applied parallel to a surface or area, and designated by the symbol τ (tau).

Strain. The response of a material to stress and indicated by the symbol σ (sigma).

Stress-trajectories. Also called *isostatics.* A system of curves related to orthogonal reference axes and whose directions at any point are the directions of principal stress axes.

Surface forces. Forces applied to the surface of a body or existing at an interface between two materials.

Tensile strength. The strength of a material under tensional stress.

Triaxial compressive strength. A measure of the strength of a material under a directed stress when the material is confined by a fluid under pressure or by an enclosing jacket of metal or other material.

Uniaxial compressive strength. A measure of the strength of a material under a directed stress.

Viscoelastic material. Material with combined elements of elastic and viscous (fluid) materials.

Voigt material. Firmo-viscous material, which see.

Viscous material. Also called *Newtonian material.* A material such as a fluid that can not withstand any shear stress without deforming permanently. The response to shear stress is a function of viscosity and time.

Young's Modulus. A constant for each elastic material which equals the ratio of the stress σ (sigma) to the strain ϵ (epsilon).

FORCE AND STRESS

Force is a vector quantity identified by a direction and an intensity. The unit of force in the C.G.S. system is the *dyne,* a force that will accelerate a one gram mass at the rate of one centimeter per squared second. *Stress* is a tensor quantity requiring for its description more than three components (commonly nine) in a coordinate reference system and is expressed as a force acting on a unit area. The force may act on an *exterior boundary surface* (including an interior opening such as a tunnel) of a body, or it may be presumed to be acting on an imaginary *interior surface.* Forces and stresses are described as *hydrostatic* (equal forces or stresses acting in all directions), *tensional, compressional, torsional,* or *coupled.* Coupled forces or stresses are opposing but not in the same line, and sometimes are called *shear* or *rotational* forces or stresses. *Body forces* act throughout a body and are produced without physical contact with other bodies. Examples are electrical, thermal, magnetic, gravitational, and inertial forces.

MECHANICAL PROPERTIES OF ROCKS

The mechanical properties of rocks measured in the laboratory or the field provide a basis for prediction of the behavior of the rocks under stress in foundation excavations for dams and other heavy structures, road cuts, surface mining excavations, and underground excavations. A variety of mechanical properties may be measured, including impact hardness, abrasion hardness, density, toughness, permeability, porosity, and the like, but the properties of most concern here are the properties measuring strength as related to the competency of rocks in underground excavations.

Three kinds of samples for testing are generally available for study and/or prediction of the strengths of rocks in subterranean excavations: (1) samples taken from surface exposures of rocks above the tunnel, (2) cored samples obtained from exploratory diamond drilling, and (3) cored samples taken from underground exposures of rock as excavation proceeds. Under many conditions it is difficult to collect unweathered samples at the surface that will truly represent rocks at depth below the weathered zone.

The strength of a rock is measured or estimated in a variety of ways, and for elastic materials is described by means of *elastic constants* which are characteristic of each material. Elastic constants are measured by static means, in which pressure is exerted on a material, and the response is measured, or by dynamic means, in which velocities of propagated energy waves are measured.

Important mechanical properties measuring strength are the following:

(1) *Uniaxial compressive strength.* Determined by uniaxially compressing a cylinder or prism of rock to the point of failure.

(2) *Tensile strength.* Determined by exerting tension in the direction of the axis of a cylinder to failure or by applying compression at a point to a cylinder lying on its side (the Brazilian test).

(3) *Flexural strength.* Determined by supporting a beam at its ends and loading the midpoint of the beam to failure. This procedure creates tension on the under side of the beam.

(4) *Unconfined shear strength.* Methods of determination yield inconsistent results but commonly involve placing a specimen between knife edges and exerting pressure.

(5) *Triaxial compressive and shear strength.* Determined by placing a cylinder of rock in a compression chamber filled with fluid and then applying uniaxial loading to failure. Testing permits evaluation of normal and shear stress acting on a plane of failure and the angle and coefficient of internal friction. Triaxial testing yields data of considerable value in rock mechanics, particularly with respect to rocks under load, as in a tunnel excavation. Accordingly, triaxial testing will be reviewed briefly in a subsequent section of this chapter.

STRESS–STRAIN–TIME RELATIONSHIPS

Behavior of materials under stress

Strain is the deformational response to stress. Commonly in the laboratory the stress is applied to a small cylinder of material in a single direction (uniaxial stress) and the shortening of the cylinder, the strain, is measured and expressed as a ratio or a per cent. Natural and artificial materials show a wide range of behaviors under stress, ranging from almost perfectly elastic to almost perfectly ductile. Fig. 17 shows some idealized stress–strain curves for various kinds of materials tested over short time intervals.

Depending upon behavior under stress, materials may be classified as *very brittle, brittle, moderately brittle, moderately ductile,* or *ductile.* A parallel but not equivalent terminology classifies materials as *elastic, elastoplastic, viscoelastic* and *firmoviscous,* and *viscous* (see Glossary).

Some materials undergoing testing, for example certain plastic or ductile materials, develop increasing strength after a certain interval of deformation. This phenomenon sometimes is described as *work* or *strain hardening.* This phenomenon also is observed in underground openings, especially in plastic rocks of very low initial strength, which, after a period of time, become stiffer and more resistant to strain dislocation.

Fractures in rocks

Elastic rocks stressed to the point of failure in the laboratory frequently develop

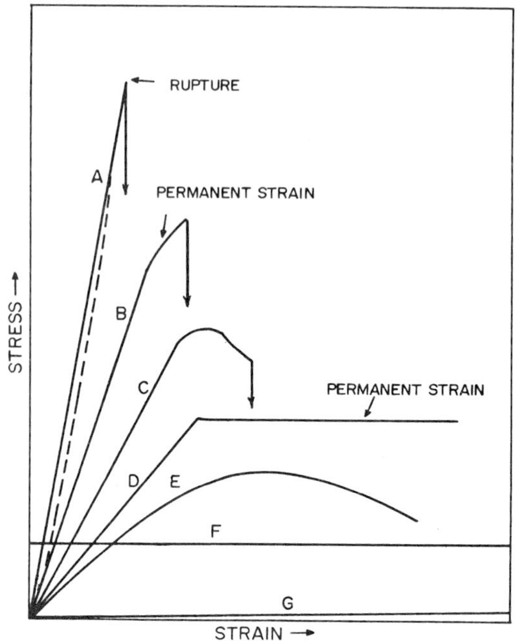

Fig. 17. Idealized stress–strain curves for several kinds of material. *A* = elastic material which obeys Hooke's law to point of rupture, dashed line suggests behavior of many elastic materials to reduction of stress before rupture point is reached; *B* = elastic material which becomes permanently strained just before rupture; *C* = material that displays ductile behavior just before rupture; *D* = material which behaves elastically, then at a critical stress becomes plastic in its behavior; *E* = ductile material; *F* = plastic material; *G* = liquid material.

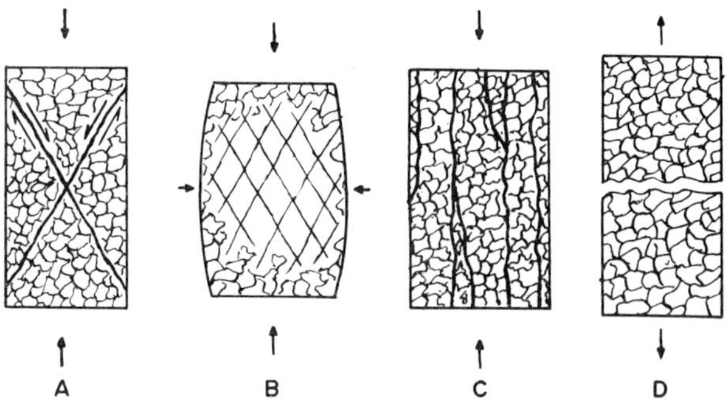

Fig. 18. Fractures developed under uniaxial stress. A. Conjugate shear fractures. B. Multiple conjugate shear fractures formed when sample is confined. C. Longitudinal tension fractures, sometimes called "lateral extension fractures". D. Extension fracture formed in tension.

fractures by mechanical disruption. Common types of fractures developed under uniaxial stress are shown in Fig. 18.

Behavior of materials under triaxial stress

Stress–strain measurements of unconfined samples yield information concerning the behavior of rocks under stress and enable estimation of strength and competence under load. However, triaxial stress–strain measurements are much more pertinent to an understanding of the behavior of rocks confined at depth by a superjacent rock load. Fig. 19 shows how dead-load weight increases with depth for water, and for rocks in the density range 2.5–3.0, as expressed in the various units in common use in materials testing. If dead-load weight is equated to hydrostatic

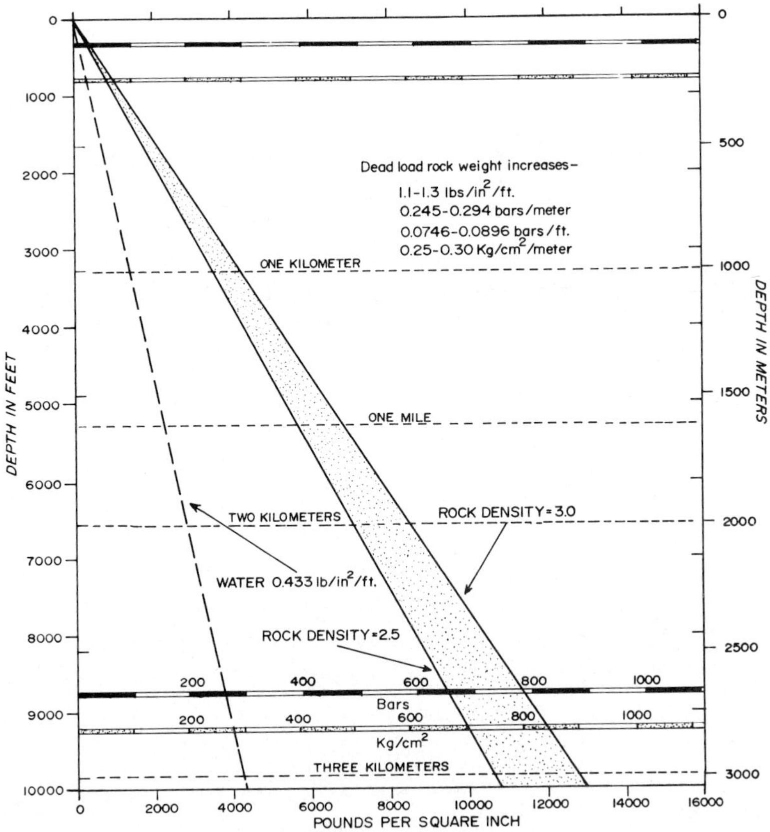

Fig. 19. Increase in dead load weight with depth of water and rocks within the density range 2.5–3.0. No allowance is made for increases of density related to compressibility or for density decrease because of elevated temperatures with increasing depth.

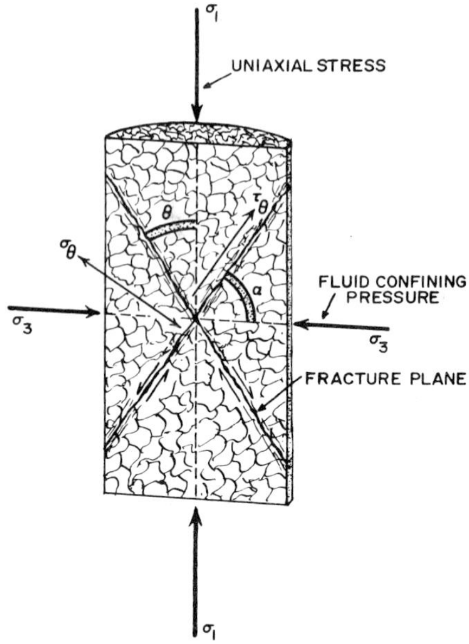

Fig. 20. Cylinder of rock under triaxial compression. Rock has failed along conjugate fractures which make and angle θ with the direction of uniaxial compression, σ_1; σ_θ is the stress normal to a fracture and τ_θ is the shearing stress parallel to the fracture.

pressure, it is seen that rock-load pressures at even moderate depths must be an important factor in the behavior of rocks under stress.

Fig. 20 shows a rock cylinder under triaxial stress and also indicates some of the notations used in identifying conjugate fractures developed by failure.

A very useful approach to a rationalized understanding of the behavior of rocks under triaxial stress is provided by the use of theory developed by Mohr and subsequently reviewed and elaborated by Balmer (1952), Obert and Duvall (1967), and others.

Let σ_1 and σ_3 be the uniaxial stress and the confining stress, respectively, during triaxial compression testing (Fig. 20). The normal stress, σ_θ, and the shear stress, τ_θ, on a plane inclined at an angle θ to the direction of uniaxial stress, σ_1, can be estimated by the following equations, given without proof:

$$\sigma_\theta = \tfrac{1}{2}(\sigma_1 + \sigma_3) - \tfrac{1}{2}(\sigma_1 - \sigma_3)\cos 2\theta \qquad (4\text{-}1)$$

$$\tau_\theta = \tfrac{1}{2}(\sigma_1 - \sigma_3)\sin 2\theta \qquad (4\text{-}2)$$

The total shearing resistance, τ, of an isotropic substance to failure, according to theory, is the sum of the so-called *cohesive strength*, τ_0, a parameter that can not be measured directly and a term that represents frictional resistance to displacement along a potential plane of failure. This term is the product of the effective normal stress, σ_θ, across the plane and the so-called *coefficient of internal friction,* which is expressed as $n = \tan\phi$, where ϕ is an angle corresponding to the angle given consideration in ordinary sliding friction.

Accordingly:

$$\tau = \tau_0 + \sigma_\theta \tan\phi \tag{4-3}$$

Of interest is the angle θ for which the shear stress τ_θ, is at maximum and the normal pressure, σ_θ, is at a minimum. By appropriate substitution of eq. 4-1 and 4-2 into eq. 4-3, we arrive at:

$$\theta = \pm 45^\circ \mp \tfrac{1}{2}\phi \tag{4-4}$$

Potential planes of failure exist in conjugate pairs, and the acute angle between them is bisected by the direction of σ_1, the uniaxial stress. Because the fluid confining pressure, σ_3, is exerted uniformly on the sides of the rock cylinder during triaxial compression, failure generally results in the development of cones of failure rather than a single set of conjugate planes, and the angle θ is in fact half the apical angle of a cone.

Conical shear failure is rarely observed in actual rock bodies in the field. Instead, two sets of intersecting conjugate fracture planes are common. It must be realized, accordingly, that the conditions of triaxial compression in laboratory tests are rarely duplicated in natural rock masses. However, if it is assumed that there are three mutually perpendicular directions of stress, one of maximum stress, σ_1, one of intermediate stress, σ_2, and one of minimum stress, σ_3, the two sets of conjugate planar fractures can be considered as intersecting in a direction parallel to the direction of intermediate stress application. In triaxial testing in the laboratory, because of the conditions imposed by the testing procedure, σ_2 must be considered as equal to σ_3.

Eq. 4-1—4-4 are the basis for construction of Mohr's stress circles, which enable graphical solutions for the values of τ_0 and ϕ (Fig. 21). With data derived from a succession of triaxial compression tests, in which the confining pressure, σ_3, is varied, envelopes for ultimate strength of a material may be drawn as in Fig. 21. Although the envelopes are generally somewhat curved, indicating some variation in ϕ as a function of confining pressure, it is common practice to determine ϕ and τ_0 by using the best straight-line fit for an envelope. If the convention is followed that compressive stresses are negative, the abscissa for normal stress should be construed to increase in a negative sense to the right in Fig. 21.

Mohr's theory of failure is based on the premise that a material will fracture or

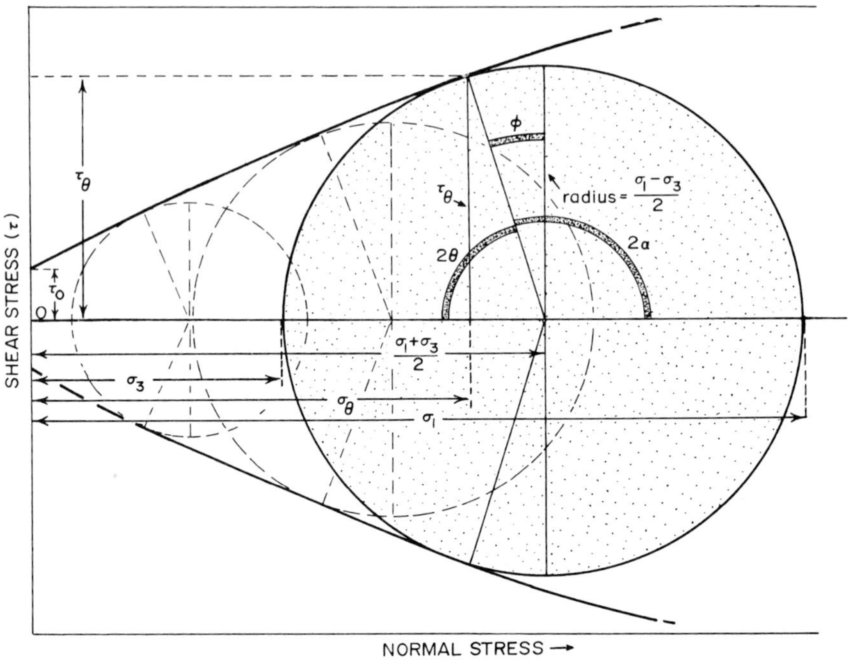

Fig. 21. Mohr construction for ultimate strength of a rock. A Mohr stress circle (stippled) with radius $(\sigma_1 - \sigma_3)/2$ and a center at $(\sigma_1 + \sigma_3)/2$ correlates with a normal stress of σ_θ (eq. 4-1) and a shear stress of τ_θ (eq. 4-2) on planes inclined at angles $\pm 2\theta$ to the direction of σ_1. The Mohr envelopes are tangent to a series of circles obtained from a succession of triaxial tests and yield the cohesive strength τ_0 at zero normal stress and the coefficient of internal friction n = tan ϕ from the average slope of the envelope (eq. 4-3). At points of tangency $\theta = \pm 45° \mp /2$ (eq. 4-4). Curvature in the envelopes indicates that tan ϕ is not necessarily constant, although, in general practice, tan ϕ is assumed to be a constant based on a straight-line envelope.

will begin to deform permanently when the shear stress τ_θ on a fracture plane or slip plane has increased to a value depending on the normal stress acting on the same plane. For any state of stress represented by a circle within the tangent envelopes a material will not fail, but, if any part of a stress circle lies outside of an envelope, the assumption is that critical stresses will have been exceeded, resulting in failure. Thus the envelopes identify ultimate strength of a material as a function of varying stresses.

The presence of fractures and other deformational features in the earth's crust is a clear indication that the rocks failed in some manner in response to various kinds of stresses acting sometime in the geologic past, and when the failure occurred at depth, the rocks were totally confined. A tunneling operation abruptly changes boundary conditions and may cause dramatic changes in the directions and magnitudes of stresses existing in confined rocks prior to excavation. Under these circum-

TABLE XI

Representative stress–strain data for some rocks and unconsolidated materials subjected to triaxial testing

	Temperature (°C)	*Confining pressure* (bars)	*Ultimate strength* (bars)	*Fault angle* (θ) (degr.)
Igneous rocks	150	1,010	3,290–8,000	24–33
Sedimentary rocks				
sand	24	490	971	–
sand	24	1,010	1,820–2,340	–
sandstone	24	0	100–1,870	10–19
	24	1,010	2,290–6,870	14–34
siltstone	24	0	280– 480	20–24
	24	1,010	1,790–2,260	–
clay (air dried)	24	0	10	0
	24	920	930	–
shale	24	0	390– 740	5–10
	24	1,010	580–2,450	10–33
limestone	24	0	380–3,470	10–40
	24	1,010	2,060–5,990	24–30
Metamorphic rocks	24	0	980–3,590	10–31
	24	1,010	2,960–10,790	29–45

stances the theory and data of triaxial testing in general cannot be translated quantitatively into practical solutions of many kinds of problems encountered in design and construction of tunnels, especially tunnels in fractured or highly-anisotropic rocks. Nevertheless, the determination of ultimate strengths of rocks by the methods of triaxial testing provides at least a semiquantitative basis for estimation of the inherent competency of rocks that might be encountered during tunnel excavation,

TABLE XII

Representative values taken from Mohr plots for rocks subjected to triaxial testing at low confining pressures (loaded normal to bedding or layering, if present)

	Crushing strength (bars)	*Cohesive strength* (bars)	*Coefficient of internal friction* ($\tan \phi$)
Igneous rocks	720–1730	140–320	1.0–2.1
Sedimentary rocks			
sandstone	300–1060	60–180	1.0–1.6
shale and siltstone	15–1230	3–240	0.8–2.1
limestone	800–1330	130–360	0.7–1.7
Metamorphic rocks	70–1360	20–260	0.5–2.9

especially when such rocks are not intersected by fractures developed at some previous time of strain failure.

Table XI and XII give some values obtained from triaxial testing for a representative group of rocks. The values were selected from a comprehensive summary by Handin (1966).

Inspection of Table XI and XII brings out certain conclusions of interest. (1) The competency of rocks differs widely even within rocks of the same general petrographic classification. (2) Ultimate strength increases notably with an increase in confining pressure. (3) The shear angle θ tends to increase with increased confining pressure, and for many materials is in the range 10–33 degrees. A calculated average value is near 30 degrees. (4) The tests yield results that are in complete accord with practical experience derived from field excavations in a variety of natural materials and from knowledge of rocks obtained in laboratory petrographic examinations.

Effects of temperature on strength of rocks

It is well known that heating, particularly in the higher ranges of temperature, lowers the strength of rocks. However, within the temperature range of tunneling operations it may be considered that temperature effects on the strength of rocks are negligible.

Effects of time on the strength of rocks

Materials subjected to stresses of long duration generally fall well below yield or rupture strengths that are determined in typical short-term laboratory tests. The strength of rocks subjected to continuing stresses over a period of a few hours to many years in an underground opening are of great concern to the engineer and the geologist, and it is unfortunate that only a few studies of time-dependent strength of rocks have been made of actual examples in the field or in the laboratory. Sufficient investigation has been completed, however, to enable certain generalizations such as those implicit in Fig. 22, a plot of strain against time assuming constant differential stress and initial short-term elastic behavior.

In an underground excavation creation of the opening typically results first in an instantaneous elastic strain as, for example, at the moment of detonation of an explosive charge. Following shortly upon the instantaneous strain a period of transient strain of relatively short duration may be observed, commonly lasting only a few minutes to a few hours, and in some instances accompanied by local failure of rocks immediately adjacent to the opening. Rock bursts may develop during transient strain, or at a later time. Steady state strain occurs at a more-or-less uniform rate over a moderately to very long time, measured in days, months, or even years.

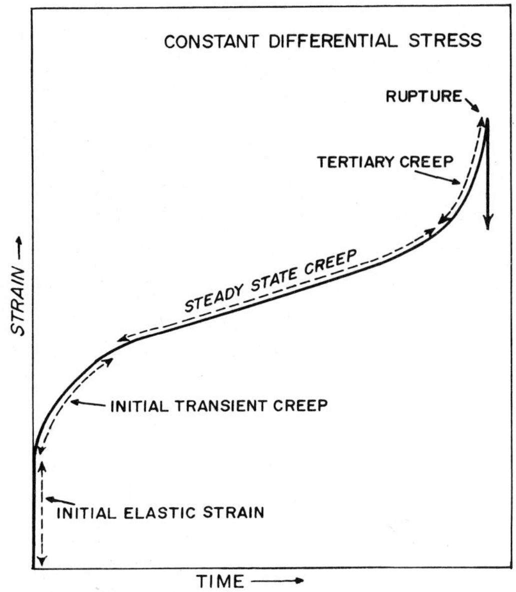

Fig. 22. Idealized strain–time curve for material under constant differential stress. No scale.

In many openings rock failure after a considerable time interval of steady state creep is preceded by a marked increase rate of strain over a short interval of time, identifying an interval of *tertiary creep*. Commonly, tertiary creep is characterized by separation of small rock fragments and "bumps", sharp rock shocks and noises resulting from quick, slight dislocations of rocks along old or newly created fracture surfaces.

Rock-mechanics instrumentation in underground openings serves one of its most useful functions in the study of creep of rocks, that is, strain with time, in an effort to preserve the integrity and continued safety of an opening.

Tertiary creep is not always observed, suggesting that steady-state creep may terminate in abrupt failure of rocks without any prior warning, or, as an alternative interpretation, the time interval for tertiary creep is so short that the time–strain curve becomes essentially vertical.

Effects of creep in underground openings have been noted especially in rock pillars left to support multiple openings in mining operations in layered deposits. Typically, if the pillars are not of sufficient dimensions and spacing to support the weight of the overlying rock, failure occurs over a period of time by intermittent separation of flat slabs and curved plates from the sides of a pillar and, in some examples, by development of conjugate shears within the pillar. Similar effects have been noted in mining operations at and near the juncture of two tunnels intersecting at a small angle.

STRESSES IN UNDERGROUND OPENINGS IN ISOTROPIC ELASTIC ROCKS

An appreciation of the intensities and distribution of the stresses that are associated with undergound openings is obtained by an application of elastic theory to two-dimensional shapes of simple geometry in isotropic rocks. Anisotropy in rocks and fractures of any kind greatly complicate stress analysis, but an understanding of the distribution of stresses under idealized conditions in elastic rocks provides a basis for meaningful subjective extrapolation into more complex situations.

Certain conventions are observed in the subsequent discussion. Compressive stress has a negative sign and tensile stress has a positive sign. The stress at or near the boundary of a tunnel opening is conveniently expressed as a *stress concentration*. A *positive* stress concentration at a point means that the stress has the same sign as the applied stress. Conversely, a *negative* stress concentration differs in sign from the applied stress.

The simplest way to analyze stress distribution in tunnel openings at depth is to make assumptions regarding the relative magnitudes of the vertical stress, depending on the load of overlying rock, and the horizontal stress exerted in a horizontal plane. In Fig. 23 three assumptions are made regarding the vertical load stress, S_v, and the horizontal stress, S_h. In Fig. 23A a tunnel near the surface is assumed to have no horizontal component of stress. In Fig. 23B, typical of many tunneling operations at moderate depth, say of the order of one mile, it is assumed that the horizontal stress is equal to one-third of the vertical load stress. Finally, in Fig. 23C the tunnel is located at such great depth that $S_h = S_v$, a condition that is realized when stresses are hydrostatic or the rock behaves plastically.

Stress concentrations at the interface between a tunnel opening and elastic rock are shown in two geometrically simple cross-sections in Fig. 24. In Fig. 24A a

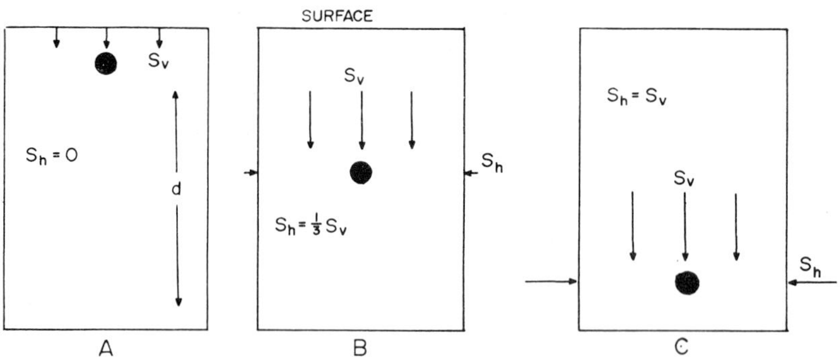

Fig. 23. Diagrams showing basis for making assumptions regarding horizontal stress (S_h) and vertical stress (S_v) in tunnel openings at various depths (d) below the earth's surface. A. At shallow depths; S_h = 0. B. At intermediate depths; S_h = 1/3 S_v. C. At great depths or plastic or semiplastic behavior of rocks; S_h = S_v.

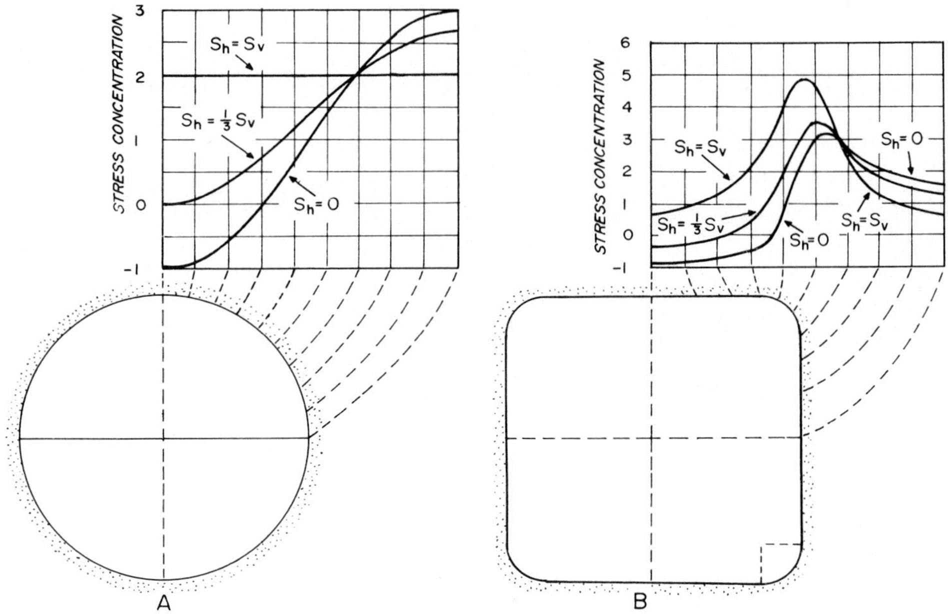

Fig. 24. Calculated stress concentrations in two dimensions for idealized tunnel cross-sections. A. Circular cross-section. B. Square cross-section with rounded corners.

circular opening is shown, and curves are drawn for various assumptions regarding the relationships between vertical and horizontal stresses (Fig. 23). A negative stress concentration for $S_h = 0$ indicates tension and positive concentrations for other assumptions indicate compression. Fig. 24B shows a square tunnel cross-section with rounded corners. The corners are rounded because, if the tunnel ceiling and the wall intersect at a point, the stress concentration at the point approaches infinity, and the stress analysis is not pertinent to actual tunnel openings. Again negative stress concentrations indicate tension, and positive stress concentrations indicate compression. Note the high stress concentrations in the rounded corners, as compared with the flat roof and sides of the opening.

Stress analyses have been made of tunnel cross-sections of a variety of geometrical shapes (Obert and Duvall, 1967; Jaeger and Cook, 1969), but, without regard to shapes of openings, certain generalizations may be made. These are:

(1) Stress concentration decreases rapidly with distance from an opening. The distance is of the order of a few feet to several feet, depending on the properties of the rocks.

(2) As the horizontal component of stress increases relative to the vertical stress, corresponding to a change from uniaxial stress to hydrostatic stress, the boundary stress on the horizontal axis decreases.

(3) The stress distribution on the periphery of a tunnel opening is dependent on the shape, but not the size of the opening.

(4) Tunnel openings with sharp corners should be avoided because stress concentrations are associated with such corners.

(5) For a hydrostatic stress field ($S_h = S_v$), as in viscous and plastic materials (squeezing ground) a circular tunnel section is preferred, in part because no tensile stresses develop (see Fig. 24A).

ISOTROPISM AND ANISOTROPISM IN ROCKS

Laboratory testing of the strength of rocks generally utilizes cylinders of unfractured rock, and, in some tests, attention is paid to rock layering in orienting test cylinders. Testing of layered samples generally is conducted on cylinders oriented parallel to and at right angles to the layering, and samples commonly show much greater compressive strength in cylinders cored at right angles to layering than in cylinders oriented parallel to layering. When planar features are oriented 15–60° to the direction of maximum principal compression, fractures tend to develop at failure parallel to the planar elements, and failure occurs in stress ranges considerably below those for comparable isotropic materials.

The behavior of rocks during and after excavation in a tunnel depends to a large extent on whether the rocks are isotropic or anisotropic. In terms of their distribution and intensities the stresses that exist or are created around a tunnel opening generally are related in a complex way to rock anisotropism.

Rock isotropism or anisotropism is an expression of *fabric* as exhibited in mineral grain sizes, grain shapes, orientation of nonequidimensional grains, articulation of the grains and the amounts and space distribution of the various mineral components. Fabrics that require characterization by study under the microscope are *microfabrics* or *microscopic fabrics*. Fabrics that can be observed in hand specimens, laboratory test samples, or in surfaces having approximately the same dimensions as those encountered in surface or underground excavations are identified as *megascopic fabrics*, and are of particular interest in the present discussion. Fabrics determined by geologic mapping of large areas of the earth's surface are *regional fabrics*. Fabrics developed by a given combination of geologic processes commonly display a characteristic symmetry, regardless of scale.

The fabric of a particular rock provides a record of the natural processes that determined the identity of the rock as expressed in its mineral composition and the three-dimensional interrelationships of the mineral components. In igneous rocks fabrics are the consequence of partial to complete crystallization of magmas under a variety of conditions within the earth's crust or on the earth's surface. Fabrics in sedimentary rock result from sedimentary deposition followed by compaction and

TABLE XIII

Megascopic fabrics in rocks

I. *Isotropic fabrics.* Megascopically visible linear or planar fabric elements are absent.
Examples:
  a. Igneous rocks crystallized from magmas which were not actively moving at the time of crystallization.
  b. Lithified pyroclastic deposits resulting from rapid deposition of equidimensional particles of exploded materials.
  c. Clastic sedimentary rocks which were formed by rapid deposition from water or air. Included are some graywackes and arkoses.
  d. Monomineralic clastic sedimentary rocks consisting of essentially equidimensional mineral grains.
  e. Sedimentary rocks such as gypsum, anhydrite, and rock salts, also some limestones, which partly or completely recrystallized from isotropic accumulations of crystals and/or fragments.
  f. Metamorphic rocks such as quartzite, marble, and hornfels which recrystallized without notable deformation from originally isotropic mineral aggregates.

II. *Anisotropic fabrics.*
  A. Linear fabrics.
    1. Monomineralic rocks.
    Examples:
      a. Metamorphic rocks consisting dominantly of prismatic or acicular minerals such as hornblende and sillimanite in parallel to subparallel alignment. *Pencil schist.*
    2. Polymineralic rocks.
    Examples:
      a. Igneous rocks containing subparallel to parallel prismatic or acicular minerals (especially hornblende) associated with other minerals displaying an essentially random (isotropic) arrangement.
      b. Metamorphic rocks containing prismatic or acicular minerals in parallel or subparallel orientation associated with other minerals in random arrangement. *Pencil gneiss.*
  B. Simple planar fabrics. Planar elements are essentially flat, and not folded or corrugated.
    1. Monomineralic fabrics in rocks with uniform grain size, but mineral displays platy, micaceous or tabular habits.
    Examples:
      a. Sedimentary accumulations of a clay mineral such as shale consisting primarily of particles of a single clay mineral such as illite or kaolinite in parallel orientation producing pronounced fissility.
      b. Metamorphic rocks consisting dominantly of a single platy or micaceous mineral such as mica or tabular feldspar in parallel to subparallel orientation.
      c. Igneous rocks, especially anorthosite, consisting essentially of tabular plagioclase feldspars in parallel to subparallel arrangement.
    2. Monomineralic fabrics in rocks with particles of unequal sizes; alternating layers contain concentrations of contrasting size.
    Examples:
      a. Quartz sandstone with finer-grained layers alternating with coarser-grained layers.
      b. Limestone with alternating layers of calcite grains with different grain sizes.
      c. Metamorphic rocks, such as quartzites or marbles, with layers containing grains of unequal size.

TABLE XIII (continued)

---

3. Polymineralic fabrics. Layers contain particles of different mineralogy and/or different grain size
   Examples:
   a. Stratified sedimentary rocks.
   b. Igneous rocks with flow layering.
   c. Metamorphic rocks, especially *banded gneiss* and *augen gneiss.*

C. Intersecting planar fabrics. Two or more planar elements of fabric are present and intersect at any angle. Planar elements may be identical (conjugate planar fabric) or may be characterized by different mineralogy and grain size.
   Examples:
   a. Metamorphic rocks which have crystallized from rocks that were subjected to shear, tensional, or torsional stress producing planar dislocations concomitant with recrystallization.

D. Two or more directions of planar fabrics and one group of planar elements is truncated by or terminates against another group.
   Examples:
   a. Some cross-bedded sedimentary rocks as *cross-bedded sandstone.*
   b. Some metamorphic rocks in which diagonal planar elements are confined to layers with a compositional identity different from adjacent layers.

E. Folded or curved planar fabrics. Mineralogic or grain size segregations are in more or less complexly folded layers or were deposited in curved or corrugated layers.
   Examples:
   a. More or less intricately folded metamorphic rocks, both schists and gneisses. Very common in high grade metamorphic rocks.
   b. Lava flows or glassy to very fine-grained igneous intrusions which moved as highly viscous masses immediately prior to consolidation and developed contorted flow layers.
   c. Sedimentary layers exhibiting curved or corrugated surfaces as in *ripple-marked siltstone*, some *cross-bedded sandstones*, and some calcareous deposits.

F. Composite fabrics. Contain various combinations of linear, planar, and isotropic fabrics. Frequently linear elements lie in one or more planar features.

---

lithification in a wide range of natural environments. Finally, metamorphic rocks have fabrics which result from recrystallization of pre-existing rocks at moderate to high temperatures and/or low to high directed forces. Fabrics in metamorphic rocks inherited from previously anisotropic rocks are *relict fabrics.* An example is *relict bedding* in a metamorphosed sedimentary rock.

In megascopically *isotropic rocks* fabric elements, especially mineral grains, have random orientations, and the physical properties of a rock are the same in all directions. There are no visible planar or linear elements of fabric. In megascopically *anisotropic rocks* linear or planar elements of fabric are present, and physical properties of a rock visibly differ with different directions.

A classification of megascopic rock fabrics, with examples, is given in Table XIII. The reader, upon consideration, will be able to identify a large number of additional examples.

STRENGTH OF ANISOTROPIC ROCKS

Parallel to subparallel arrangement of platy or micaceous minerals renders rocks susceptible to easy splitting or fracture in the direction of parallelism. The clay minerals, micas, and chlorite have a perfect mineral cleavage parallel to plates, and the tensile strength normal to the cleavage is minimal. Parallel segregations of these minerals lead to well-developed *fissility* or *rock cleavage.*

Some selected examples of planar fabrics in rocks that display fissility or rock cleavage are shown in Fig. 25.

In the direction of planar elements of fabric the tensile strength usually greatly exceeds the tensile strength normal to the direction of parallelism. Conversely, crushing strength in uniaxial compression is much greater when compression is exerted normal to a planar fabric than when the compression is exerted parallel to the fabric.

Strongly anisotropic rock fabrics present kinds of problems during excavation that are not apparent in excavation and support of isotropic or weakly anisotropic rocks. As a general rule, conspicuously anisotropic rocks, especially those consisting of alternating layers of competent and relatively incompetent materials, should be considered as having no more bulk strength on the average than the strength of the weakest layers that are exposed in an excavation. The bulk or mass strength of rocks outside of the excavation, where weak layers alternate with strong layers, is another matter, and a knowledge of the mass strength of the rocks extending for some distance above the tunnel arch or into the walls of the tunnel frequently is of utmost concern. The total mass strength of a succession of strong and weak layers (or joints) is a composite of the mass strength of each of the units of strong rock and the mass strength of each of the weak layers.

Engineering design of underground openings in strongly anisotropic rocks based solely on laboratory tests of specimens can not be successful because there is no practical theoretical or mathematical model that can take into account all of the variables that exist in most natural anisotropic rock bodies. In addition, rock-mechanics instrumentation frequently is not feasible for economic reasons, or simply because there is not sufficient time to make a thorough investigation of stresses and strains in rocks, as for example, in the situation where concrete lining is placed immediately upon completion of a tunnel or heavy supports must be installed to contain the rock immediately after excavation.

An empirical approach to estimation of mass strengths of layered rocks in underground openings has been made by Protodyakonov (1964). This approach is based on a comparison of massive properties of various kinds of layered rocks with small-scale properties of the same rocks measured in the laboratory. The formula proposed by Protodyakonov for estimation of in situ mass rock strengths is:

$$S/S_m = 1 + b(m-1)/(b+L) \qquad (4\text{-}5)$$

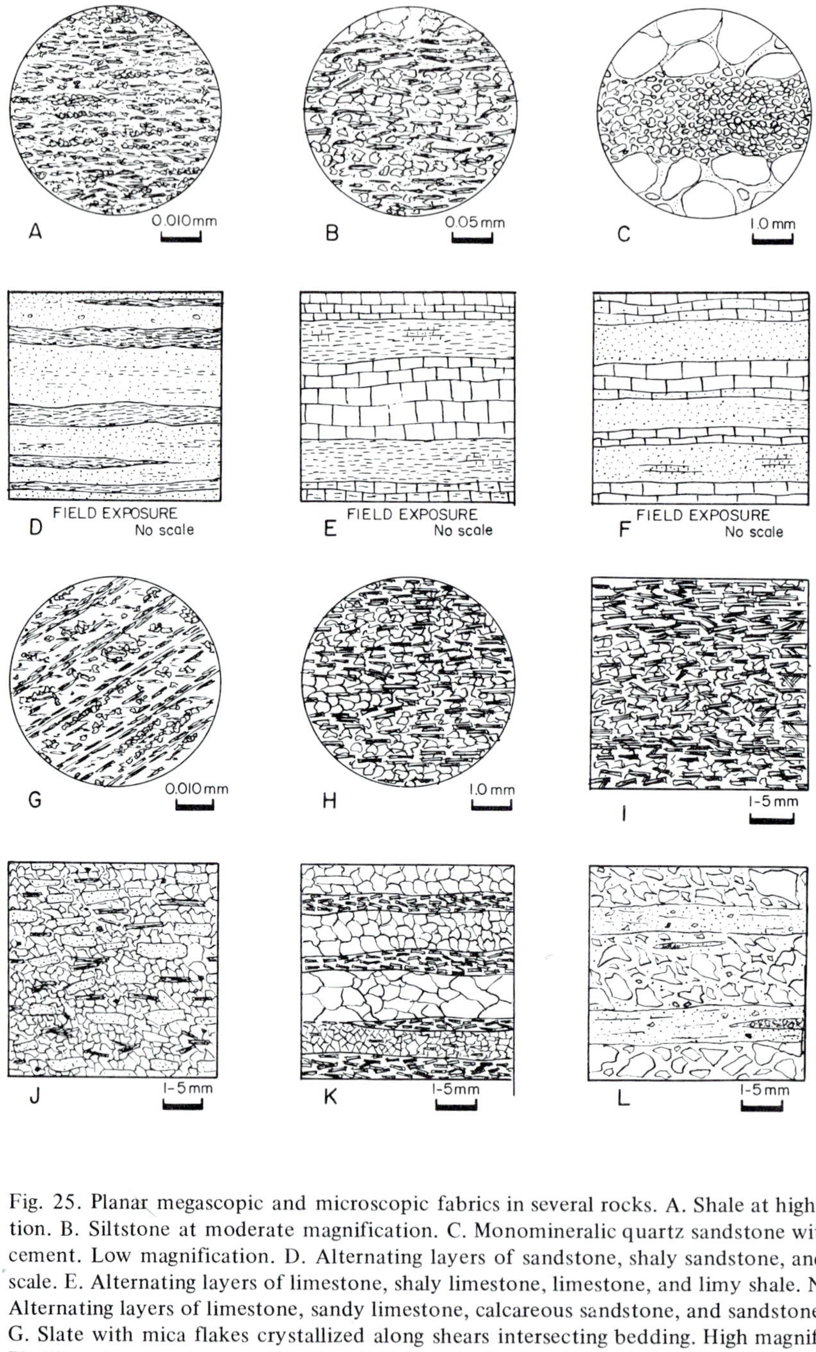

Fig. 25. Planar megascopic and microscopic fabrics in several rocks. A. Shale at high magnification. B. Siltstone at moderate magnification. C. Monomineralic quartz sandstone with siliceous cement. Low magnification. D. Alternating layers of sandstone, shaly sandstone, and shale. No scale. E. Alternating layers of limestone, shaly limestone, limestone, and limy shale. No scale. F. Alternating layers of limestone, sandy limestone, calcareous sandstone, and sandstone. No scale. G. Slate with mica flakes crystallized along shears intersecting bedding. High magnification. H. Phyllite at moderate magnification. I. Schist. J. Gneiss. K. Banded gneiss. L. Pyroclastic rock with alternating layers of breccia and tuff (volcanic ash).

in which S is the tensile or compressive strength measured in a test core in the laboratory, preferably under triaxial compression with an appropriate confining pressure. S_m is the equivalent strength of the rock layer or body underground; m is the empirical constant called the *mass fracture coefficient* having a value between 1 and 10. A low number is correlated with rocks of high compressive strength such as fresh igneous rocks. A high number, say 10, might be assigned to a weak sandstone or shale with a compressive strength of 1000 bars or less. The value of m for tensile strength is approximately double the value for compressive strength and m is assigned in number between 1 and 20. L is the dimension of the laboratory test sample, usually its diameter; b is the spacing between discontinuities such as bedding planes or joints observed in cores and ranges from a fraction of an inch to several feet.

Deshwar (1970) tested Protodyakonov's equation by coring vertical and horizontal holes and installing strain measurement instrumentation in a succession of stratified sedimentary rocks associated with a coal layer. Using mass strengths calculated from the equation he found a good correlation between calculated strains and actual strains measured in the bore holes. It is clear from Deshwar's analysis, and not at all surprising, that strengths measured in the laboratory are generally considerably higher than actual mass strengths within layered rock masses underground.

Strongly anisotropic rocks because of planes of weakness commonly produce irregular cross-sections when excavated by conventional methods (drilling and blasting), and overbreak, outside of the designed limits for rock excavation, may become a problem, especially when it is planned to line the tunnel with concrete as in a pressure tunnel. Unanticipated overbreak may require significantly greater amounts of concrete in the lining than anticipated in original estimates. Overbreak also produces special problems in placing temporary and permanent supports when they appear to be needed, because of difficulty in estimating and compensating for the short-term or long-term stress distributions in irregular openings.

The particular aspect that a strongly layered rock presents in a tunnel heading depends on the attitude of the layering relative to the direction of the tunnel and on the nature of the anisotropism. With respect to the direction of the tunnel, layering may have any inclination from horizontal to vertical and may have any horizontal direction from parallel to right angles to the course of the tunnel.

Ease of tunnel excavation, as measured by cost and rate of progress, whether by conventional methods, or by means of tunneling machines (moles), can be expected to differ with the attitude of layering relative to the direction of the tunnel. Generally, because of a greater tensile strength parallel to layering than normal to the layering, tunneling is more difficult parallel to the layering than perpendicular to layering. Tunneling difficulty in obliquely oriented layering generally lies somewhere between the two extremes.

Fig. 26. Overbreak by collapse of bedded, incompetent mudstone. Roberts Tunnel, Colorado. (Photo by Denver Board of Water Commissioners.)

Examples of overbreak in the Roberts Tunnel, Colorado, are given in Fig. 26 and 27 and indicate the strong influence of anisotropism on the shape of a tunnel opening excavated by drilling and blasting. The examples selected are typical and do not represent extreme conditions in the tunnel.

A noteworthy feature of the tunnel cross-sections is the tendency for the average tunnel opening in layered rocks to develop an elliptical shape by overbreak. The longer axis of the elliptical shape is perpendicular to the apparent dip of the layering, as observed in the tunnel heading. In contrast (Fig. 27C) overbreak was at a minimum in massive strong granite, and the outlines of the tunnel opening on the average conformed very closely to designed excavation limits as identified by the "pay line".

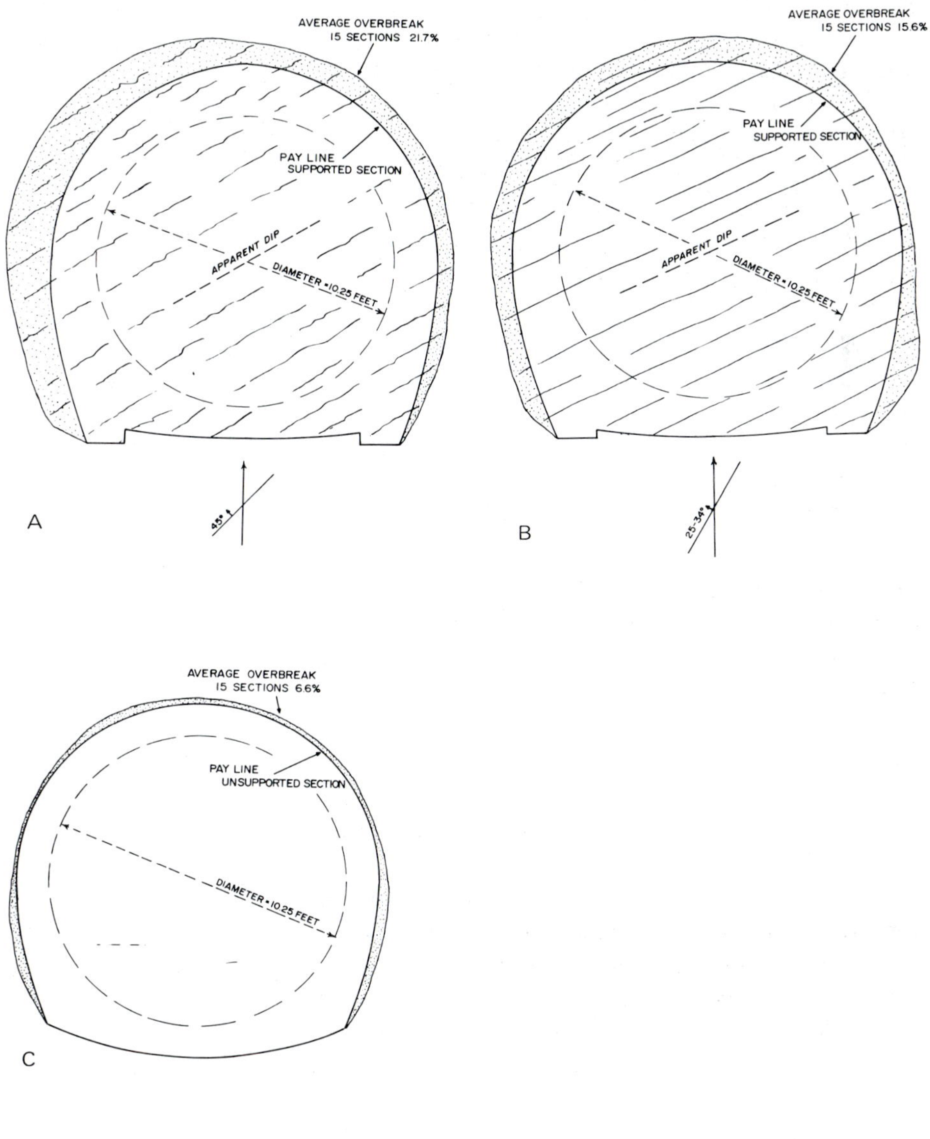

Fig. 27. Average overbreak in 15 sections in a 150-ft. interval in biotite schist and gneiss (A), in shale (B) and in massive granite (C), in the Roberts Tunnel, Colorado.

MEASUREMENT OF STRESS IN UNDERGROUND OPENINGS

The calculation of stress intensities and distributions in underground openings and the experimental study of stress distributions in scale models in the laboratory contribute much to the understanding of the behavior of elastic materials under the weight of overlying rocks. A most rewarding study is the in situ determination of stress in the walls and roof of tunnel openings because the results may be correlated with the characteristics of actual in-place rock bodies. The development of techniques for measuring stresses in underground openings represents a major advance in the practical application of rock mechanics. No attempt is made here to give more than passing notice to these techniques. The interested reader will find a very extensive literature with excellent summaries by Jaeger and Cook (1969), and Obert and Duvall (1967).

Stresses in rocks adjacent to underground openings are of two general kinds: (1) stresses related to the shape and depth of the opening and generated at the instant that the opening is excavated. These stresses may change in intensity and distribution with time because of adjustments due to creep or semiplastic behavior, and (2) generally unbalanced residual stresses frozen into the rock and sometimes called *tectonic stresses.* Residual stresses of large magnitude are preserved only in elastic materials.

Techniques for in situ measurement of stress are numerous and varied. Particularly significant and consistent results have been obtained with two methods: (1) the borehole deformation-overcoring method, and (2) the flatjack method.

In the borehole deformation-overcoring method a strain gage is placed in a small borehole and a reading is taken. Then a circular cut outside of and concentric with the borehole is drilled to relieve the stress around the borehole and a second reading is taken on the gage. The difference is the *borehole deformation.* Repetition of the procedure at successively greater depths in the borehole, together with appropriate rotation of the strain gage enable calculation of maximum- and minimum-stress directions and concentrations.

In some particularly strong elastic rocks the borehole-overcoring method does not relieve residual tectonic stresses as indicated by the fact that repeated overcoring with cuts of different radii yields a variety of results, suggesting that residual stresses may be retained even in thin shells for long periods of time.

The flatjack method involves the following steps: (1) Strain gages are placed on a rock surface oriented to measure strain in a particular direction and a reading is taken. (2) A slot is then cut in the rock between the gages to relieve the stress, and a flatjack, a circular device capable of being expanded under pressure, is grouted into the slot. (3) Pressure is applied to the flatjack until the readings on the gages equal the original readings. The pressure exerted by the flatjack is considered as being equal to the stress normal to the flatjack before the slot was cut.

Contrary to predictions of theory, maximum-stress concentrations do not always occur at the rock-tunnel interface, but may be located from a few inches to several feet from the tunnel opening, an observation explained by fracturing around the tunnel excavation or nonelastic behavior of the rock. In most examples the stress concentrations strictly identified with the tunnel opening disappear within distances measured in tens of feet from the tunnel opening.

STRENGTH OF ROCKS FROM SLOPE ANALYSIS

The geological examination of the site for a proposed tunnel starts with the examination and detailed mapping of the surface geology above the tunnel. Commonly bedrock exposures of the surface are scattered, and surficial deposits conceal much of the area under study. Depending on the geomorphologic history of the area and the resistance of bedrock to erosion and weathering, there remain topographic features of various sizes, shapes, and slopes, that, to the trained observer, tell much concerning the distribution and competency of bedrocks partly to completely buried under the surface cover.

Initially developed steep slopes in competent rocks result from fault displacement producing fault scarps, rapid downward erosion by turbulent streams with high gradients, and valley glaciation which tends to produce steep-walled, flat-floored valleys. Processes that tend to reduce the steepness of existing slopes are chemical and mechanical weathering, downslope movement under the influence of gravity of the products of chemical or mechanical weathering, further erosion by wind and water, and mass failure of over-steepened slopes. A distinction should be made between slopes developed in or over bedrock by erosion and local downslope movement of products of weathering and mass failure and slopes associated with deposition of materials transported for some distance from a point of origin by water, wind, or ice, as in alluvial fans, sand dunes, and deposits of glacial till.

Competency in rocks is measured not only by their ability to maintain steep slopes, however they may have been formed, but also by their degree of resistance to abrasion or erosion. Of course, competency of initially fresh rocks may be greatly reduced by chemical alteration as discussed in Chapter 3 or by the development of multiple fractures, a subject that will be discussed in a subsequent chapter. In the present discussion attention is paid only to initial competence, without regard to the possibility of reduction of competence by alteration or fracturing.

Fig. 28 illustrates selected examples of geologic and topographic features that permit estimation of the relative competency (strength) of a variety of rock types at the surface, and presumably, at depth.

Slopes at the earth's surface may be discussed in terms of a *slope of equilibrium,* a slope that generally is composite and consists of a bedrock slope which by one

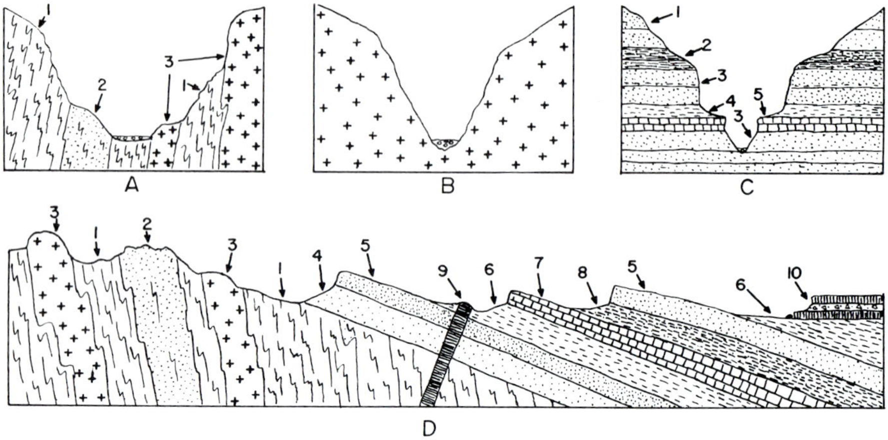

Fig. 28. Surface features and slopes developed by erosion in a variety of rock types. Surficial deposits are not shown.

A. Glaciated canyon in a mountainous region. Quartzite (*2*) and granite (*3*) have resisted glacial scouring more than layered schist and gneiss (*1*) and are judged more competent.

B. Stream-cut canyon in massive rock such as granite. Steep slopes indicate high strength in the rock.

C. Stream-cut canyon in sedimentary rocks. *1* = soft sandstone with weak cement and/or clay matrix, low competence; *2* = siltstone, weak material, low competence; *3* = hard sandstone with strong cement, highly competent; *4* = shale or mudstone, low strength, low competence; *5* = thick-bedded limestone, highly competent.

D. Stream-eroded area of igneous, metamorphic, and sedimentary rocks. Competency of rocks can be judged by resistance to erosion and steepness of slopes. *1* = layered schist and gneiss, moderately competent; *2* = hard quartzite, highly competent; *3* = granite, highly competent; *4* = sandstone, either with clay matrix or weak cement, moderately competent; *5* = hard, resistant sandstone with firm cement, highly competent; *6* = shale or mudstone, low strength, low competence; *7* = thick-bedded limestone, highly competent; *8* = siltstone, low strength, low competence; *9* = fine-grained igneous intrusion forming a cross-cutting dike, highly competent; *10* = lava flows with interlayer of soft, unconsolidated pyroclastic materials, highly competent rocks with interlayered weak material.

process or a combination of processes has reached a stable profile under existing conditions of climate, and a slope over the unconsolidated waste products derived from the bedrock and partly to completely covering the bedrock. In many areas of crystalline rocks the slope of equilibrium, both in bedrock and in fragmental waste materials makes an angle with a horizontal plane approximating 30 degrees. Slopes over soft and relatively incompetent rocks develop much flatter equilibrium slopes.

Over-steepened slopes are nonequilibrium slopes, and the persistence of such slopes with time is an excellent qualitative measure of bedrock strength. Moreover, the manner in which oversteepened slopes fail or are reduced in steepness suggests the probable behavior of rocks under stress in underground openings. Fig. 29 shows selected types of slope failure and/or retreat, and gives consideration to a variety of natural processes that tend to modify slopes.

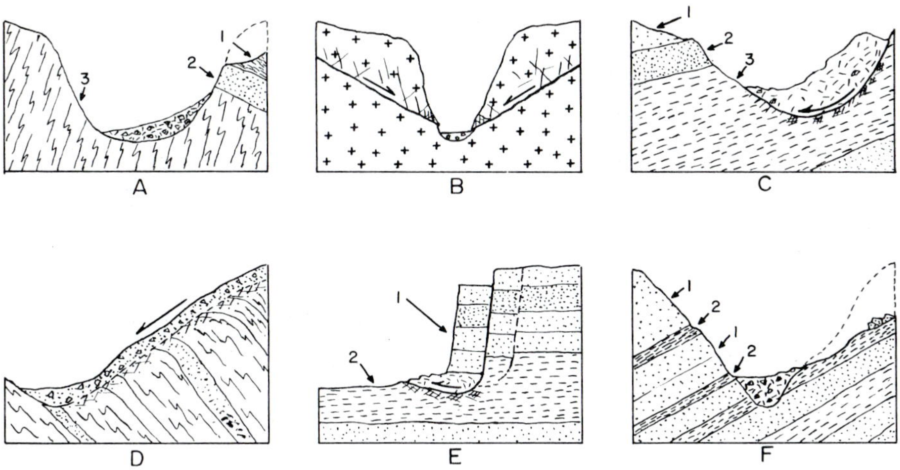

Fig. 29. Various kinds of slope modification and/or retreat in nonequilibrium slopes.
A. Steep slope in relatively incompetent shale or mudstone fails after retreat of ice in a glaciated valley. *1* = shale or mudstone; *2* = hard sandstone; *3* = schist and gneiss.
B. Oversteepened slopes in a crystalline rock such as granite result in shear failure when a critical mass has been attained through downward cutting. Many modern canyons in high–strength crystalline rocks display this phenomenon. A type of mass failure under body forces.
C. Failure by shear movement on a curved surface developed in a relatively incompetent material when a critical mass has been attained. *1* = soft sandstone; *2* = hard sandstone; *3* = shale or mudstone.
D. Downslope movement (solifluxion) of weathered materials.
E. Block settlement in competent, brittle sandstone (*1*) resting on relatively incompetent shale (2) Shale behaves plastically under load of sandstone.
F. Slope failure determined by attitude of layering in anisotropic rock. Failure occurs along a layer of weak shale (*2*) interbedded with sandstone (*1*). Layering on one side of stream-cut valley dips toward the valley. The same type of failure might involve any layered rock such as a metamorphic rock with mica-rich layers.

Qualitative estimation of the strength of rocks by slope analysis yields results in harmony with the quantitative determination of strength discussed in previous sections of this chapter.

NONEQUILIBRIUM FORCES IN ROCKS

An earth in perfect physical and chemical equilibrium with its interior, the hydrosphere, and the atmosphere, and with energy coming in from outer space, would be a monotonous globe with no appreciable surface relief and no tendency to develop unbalanced forces. Faults, folds, mountain building, and vertical and horizontal changes in the locations and attitudes of large and small surface and subsurface features prove dramatic changes in the directions and intensities of earth

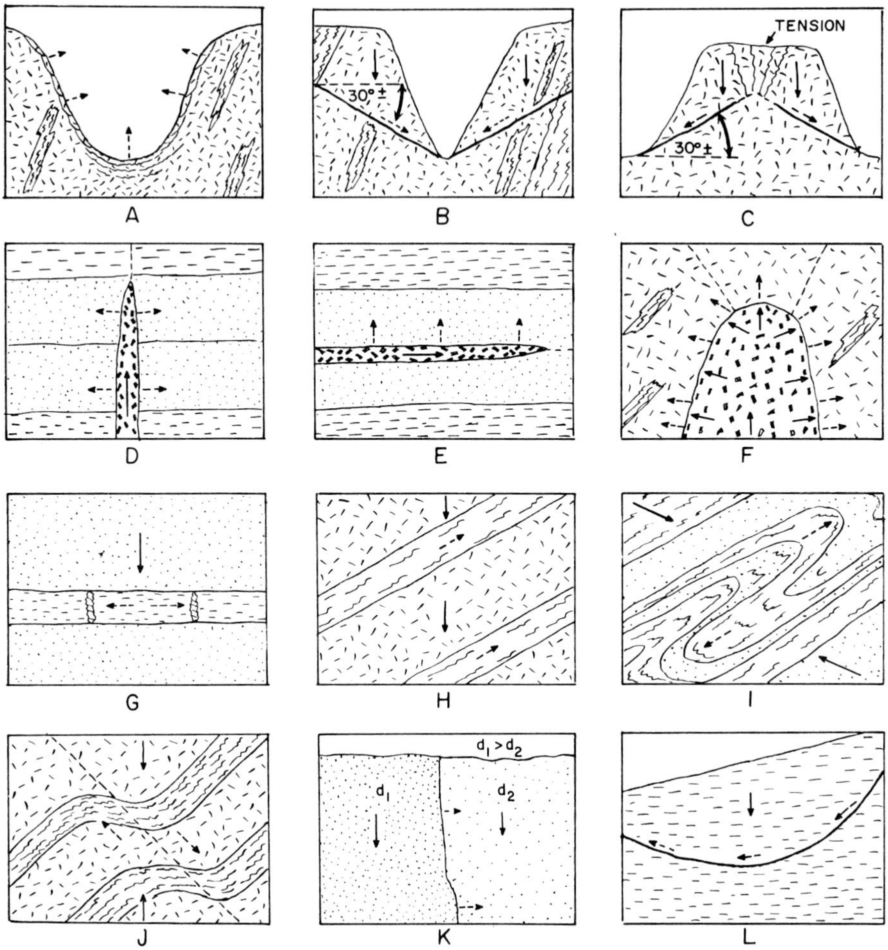

Fig. 30. Forces causing elastic or plastic dislocation or rupture of rocks under a variety of conditions. Initiating forces of maximum intensity in an unbalanced system of forces are shown as solid arrows. Resulting forces of maximum intensity are shown as dashed arrows and act in the direction of dislocation. Lengths of vectors are not drawn proportional to intensities.

A. Glaciated valley unloaded by retreat of ice. Fractures develop parallel to walls and floor by unloading associated with elastic rebound probably modified by slow thermal changes.

B. Critical accumulation of gravitative body forces resulting in shear failure of oversteepened slopes in elastic crystalline rocks. Dip of shears is near 30°.

C. Same as B.

D. Igneous dike forced into host rock causes lateral dislocation of adjacent wall rocks.

E. Igneous sill injected under pressure lifts overlying rocks. Emplacement of the sill may cause compression of adjacent rocks.

F. A stock of igneous rock generates peripheral pressures that may result in failure of the overlying country rock by fracturing (dashed lines).

G. Load exerted by overlying rock causes lateral displacement of less competent rock. Tension fractures may result.

forces in the geologic past. Earthquakes and volcanic activity prove that forces operating in the past are in existence today, and that many segments of the earth's crust are presently being subjected to unbalanced forces of large magnitude.

Analysis of the forces acting to deform or dislocate very large complex rock masses is very difficult from either a theoretical or practical standpoint. Nevertheless, much useful empirical information can be obtained by studies of rock masses in the field, and to a limited extent by study of scale models in the laboratory.

Directed or unbalanced forces are generated in rock masses by a variety of natural processes which on a large scale include:

(1) Lateral movements of segments of the earth's crust over a weak substratum (continental drift and crustal spreading).

(2) Vertical movements associated with mountain building and regional flexure of the earth's crust.

(3) Vertical movements, upward or downward, associated with isostatic adjustments in response to shifting loads at the earth's surface.

Unbalanced forces of large magnitude and wide distribution commonly are associated with the margins of continents. According to prevailing theory the semi-rigid outer portion of the earth is segmented into slabs or plates hundreds to thousands of kilometers in horizontal dimensions. These plates move with respect to each other by sliding over a less rigid underlayer. Junctures between the plates correspond with the world's active seismic belts, where concentrations of forces at shallow to great depths produce frequent rock dislocations and earthquake shocks accompanied by igneous activity. A familiar belt of seismic activity, sometimes called "the circle of fire" because of extensive volcanism, surrounds the Pacific Ocean. Clearly the locations of zones of seismic activity associated with active faults are of great concern to those planning excavations for heavy surface structures or tunnels.

Fig. 30 (continued)

H. Load exerted by overlying rocks causes movement of less competent material upward in direction of easiest pressure relief.
I. A shear or rotational force-couple causes fold deformation of semiplastic rock under high confining pressure.
J. Compressional forces produce localized elastic deformation by a shear couple. A fault may develop and release stored energy.
K. Adjacent columns of rocks with different densities generate horizontal forces, resulting in elastic or plastic dislocation at depth.
L. A gravitational load in plastic material near the earth's surface generates a curved surface of shear failure.

On a smaller scale unbalanced forces typically are generated by:

(1) Erosion at the earth's surface resulting in unequal unloading of subsurface rocks and the generation of nonequilibrium topographic features.

(2) Advance and retreat of glacial ice resulting in unequal loading followed by unloading.

(3) Localized igneous activity at depth or at the earth's surface which not only unbalances forces dependent on gravity, but also creates forces related to thermal expansion and contraction of rocks.

(4) Uneven deposition of sediments or volcanic rocks on the earth's surface.

Unbalanced forces creating minor and major dislocations of rocks in the earth's crust are not totally dissipated by being converted into work or frictional heat. The behavior of a large mass of rock subjected to stress over a period of time is not elastic, although volume elements of the rock mass may possess great strength and maximum elasticity. Accordingly, it should be expected that a reduction in intensity or the disappearance of the forces that initially caused the dislocations will leave a residue of unbalanced forces stored principally in the elastic rocks and, to a limited extent, available for release in an underground excavation.

Unbalanced forces tend to become more nearly balanced over a short period of time in fluid, viscous, and plastic materials. However, an imbalance in forces may be preserved in semielastic or elastic rock bodies over very long periods of time, and the imbalance is reduced only by rupture or subtle, long-term adjustments by creep or recrystallization. The forces retained by rocks over a period of time sometimes are called *residual forces* or *stresses.* Another expression in common use is *tectonic stress,* which is construed to mean unbalanced forces residual from a past period of rock dislocation or associated with presently active dislocation of rocks.

The assumed nature of certain kinds of forces and the direction of maximum imbalance, taken to mean the direction of maximum resulting force, are shown in Fig. 30. A more or less rapid elastic response to diminished pressure or release of pressure by failure is sometimes called *elastic rebound.* Under many circumstances elastic rebound is accompanied by rupture of rocks because of a development of critical concentration of redistributed forces. In Fig. 30 the direction of maximum imbalance in forces is taken to be the direction of maximum deformation or dislocation of rock elements. Although forces are vectorial, no attempt is made in the diagrams in Fig. 30 to indicate intensities of forces by drawing vectors with lengths proportional to forces.

ROCK BURSTS

Rock bursts are sudden, explosive separations of slabs weighing a few pounds to several hundreds of pounds from the walls or ceilings of underground openings. The bursts may occur immediately after an opening has been excavated or at any time subsequently. Rock bursts in work areas present a real threat to life, and many persons have been crushed and killed by them. Rock bursts have been noted in quarries and shallow mine workings, but it is well known that they increase in frequency and severity with an increase in depth, especially in excavations below 2,000–3,000 ft.

Explosive separation of small fragments, generally bounded by fractures in so-called "spitting ground" can be explained as the result of local concentrations of stress at or near the juncture of fractures, but separation of large slabs from apparently competent relatively unfractured rocks is not explained so easily. Common characteristics of rocks that fail by bursting are high compressive strength and brittleness. Examples are granite, quartzite, some gneisses, and very fine-grained to glassy igneous rocks, all likely to be very brittle in the unaltered state. Bursts do not develop in rocks of low strength, presumably because unbalanced stresses result in slow semiplastic adjustments so that no opportunity is provided for the local development of high stress concentrations.

Although the load of overlying rocks is an important factor in the development of rock bursts, it is clear that other forces are at work. There are many deep mine workings where bursts are not a problem. Commonly, in explaining rock bursts, an appeal is made to tectonic forces, either presently active or residual to explain the development of local concentrations of stresses sufficient to rupture rocks. It is known, in support of this concept, that residual stresses can be preserved almost indefinitely in very strong, highly elastic rocks. For example, under favorable conditions residual stresses of large magnitude can be measured in the laboratory in core samples of some high strength rocks totally removed from the original underground environment and presumably relieved of stress. It is interesting to note that some of the most severe problems with rock bursts are present in geologically very old rocks that have not been subjected to major tectonic dislocations for hundreds of millions of years. If unbalanced tectonic forces, whether active or residual, are considered to be a prime cause of rock bursts, it should be expected that rock bursts would be more frequent in areas where mountain building and crustal deformation are active now, or have been active in the recent geological past. Observation does not necessarily confirm this conjecture. By consideration of all possibilities, it becomes reasonable to assume that rock bursts are the result of the fortuitous coincidence of two or more special kinds of geologic conditions, including load as a function of depth, highly brittle rocks, residual or active tectonic forces and local geologic structure. Whatever the combination of factors may be, a sufficient force is exerted to cause explosive separation of rock slabs with long dimensions essentially parallel to the rock-opening. Several studies indicate that bursts commonly occur in rock bodies of considerable geological complexity, characterized by conspicuous rock and fracture anisotropism. The particular kind of rock anisotropism that favors

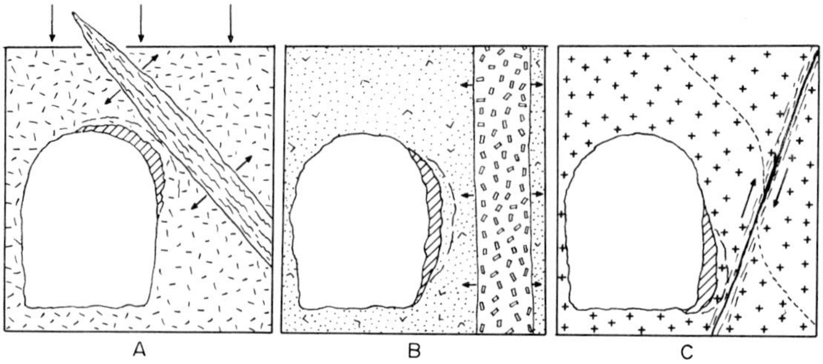

Fig. 31. Some geologic conditions creating forces resulting in rock bursts. A. Load of overlying rock in a deep tunnel in brittle granite causes semiplastic movement in a weak lens of mica-rich schist. Rock burst (hatched) occurs in tunnel at point closest to lens. B. An igneous dike which was forcefully injected into a hard quartzite is accompanied by residual stresses promoting a rock burst in the wall of the tunnel. C. Forces that have caused elastic strain in strong igneous rock are directed toward the tunnel opening to cause a rock burst.

rock bursts is one in which highly competent, brittle rocks are interlayered with less competent rocks that are capable of behaving plastically or semiplastically under unbalanced forces. In fracture anisotropy sufficient concentrations of stresses may accumulate at and near the fractures to cause rock bursts in an underground opening in the near vicinity of the fractures, whether they are faults or strong joints.

In some examples, especially in tunnels, rock bursts tend to be localized consistently on one side or another of the tunnel or in a certain segment of the tunnel arch or invert. This observation indicates that the forces causing rupture of the rock are definitely directed and are superimposed on the forces associated with the weight of the overlying rock.

Some possible conditions adjacent to underground openings that might promote rock bursts, based in part on actual examples, are shown in Fig. 31. The examples, of course, depict only a few of the great number of possible conditions, ranging from relatively simple to very complex, that might promote bursts. Prediction of rock bursts in a planned underground excavation is very difficult, even when accurate geologic projections are available, largely because of the fact that prediction of the distributions and intensities of active or residual tectonic forces is generally impossible. Nevertheless, consideration should be given to the possibility of bursts in deep openings in more or less deformed complex associations of weak and strong rocks, perhaps invaded by forcefully injected igneous intrusions.

The thought of rock bursts terrorizes miners working in underground mining operations because of the killing force of slabs of rock explosively hurled across openings. Generally the forces of large rock bursts are so great that the strongest supports cannot contain them. In some tunnels *false rock bursts* create an impression of constant danger in the minds of the miners, and, indeed, slabs of rock from false bursts are capable of crushing a person.

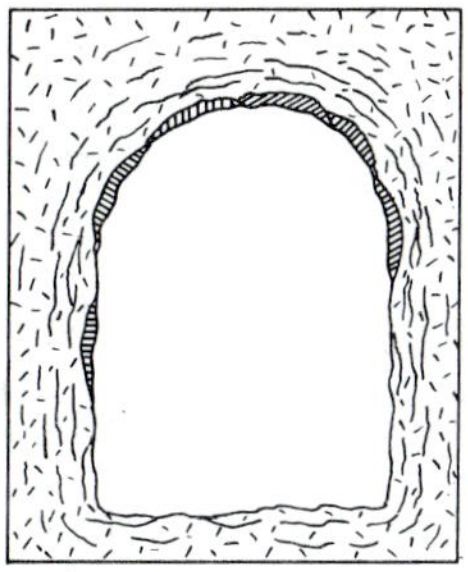

Fig. 32. Peripheral cracks caused by blasting contribute to development of false rock bursts in arch and walls of a tunnel in strong rock.

False rock bursts are observed at and close to the working heading of a tunnel and occur within a few minutes to a few hours after excavation by blasting in rock of moderate to high strength. There is little doubt that false rock bursts are initiated by blast damage to the rocks adjacent to a tunnel opening. Blast damage results in more or less irregular, continuous to discontinuous hairline fractures around the opening (Fig. 32). A delay in the separation of rock slabs, principally under the influence of gravity and from the arch and high sides of the tunnel is explained by time-dependent creep adjustments around the opening during the relatively short time of rock adjustments during the interval of initial transient creep (See Fig. 22).

False bursts may be accompanied by sharp cracking noises resulting from rapid lateral extension of blast-damage cracks at the time of separation of the rock slabs. Because of this phenomenon the miners sometimes refer to rock prone to false bursts or rapid spalling as "popping rock".

Injury from false rock bursts is most likely in continuous cycle mining operations where work is resumed at the tunnel heading immediately after drilling and blasting and before protective supports have been installed.

SQUEEZING GROUND

Rocks of very low strength, such as altered pyroclastic accumulations and poorly-lithified shales and mudstones, behave plastically and tend to move slowly into underground openings under the load of overlying rocks. At times it is very difficult to distinguish between squeezing ground and swelling ground, and a discussion of both squeezing and swelling ground is reserved for a later chapter.

REFERENCES

Balmer, G., 1952. A general analytic solution for Mohr's envelope. *Proc. Am. Soc. Test. Mater.*, 52.

Deshwar, K.H.S., 1970. Determination of in-situ strength. *Eng. Min. J.*, 171: 84–85.

Handin, J., 1966. Strength and ductility. In: S.P. Clark Jr. (Editor), *Handbook of Physical Constants. Geol. Soc. Am., Mem.*, 97: 223–289.

Jaeger, J.C. and Cook, N.G.W., 1969. *Fundamentals of Rock Mechanics.* Methuen, London, 513 pp.

Obert, L. and Duvall, W.I., 1967. *Rock Mechanics and the Design of Structures in Rocks.* Wiley, New York, N.Y., 650 pp.

Protodyakonov, M.M., 1964. Methods for evaluating of cracks and strength of rocks at depth. *Int. Conf. Strata Control Rock Mech., 4th, New York,* Addendum.

Stagg, K.G. and Zienkiewicz, O.C., 1968. *Rock Mechanics in Engineering Practice.* Wiley, New York, N.Y., 442 pp.

Talobre, J.A., 1967. *La Mechanique des Roches.* Dunod, Paris, 442 pp.

Chapter 5

FRACTURE AND FOLD ANISOTROPY IN ROCKS

INTRODUCTION

The concern in this chapter is with the processes of rock deformation by earth forces that result in modification of original isotropic or anisotropic fabrics in rocks. Deformational features that are of particular importance in tunneling operations are given particular attention, especially those features that tend to reduce the competency of rocks and/or to localize the underground circulation of water and gases.

Forces causing deformation involve rock masses of regional or local dimensions, and affect each of the mineral particles that comprise the rock masses. Deformation on a megascopic scale is of two general kinds: (1) deformation without loss of cohesion of the volume elements of a rock mass, sometimes called *flow deformation,* and (2) deformation by rupture, so-called *fracture deformation.*

DEFORMATION OF MINERALS UNDER STRESS

Forces exerted on mineral grains in rocks cause a variety of physico-chemical changes. The response to the forces in a particular mineral depends on its chemical composition, crystal structure, and the direction and magnitude of the forces. Residual or tectonic stresses stored in rocks are mainly concentrated in elastically strained mineral grains and along the surfaces of contact between the grains, and unbalanced tectonic forces in elastic or quasi-elastic rocks encountered in underground openings are the sum total of the stresses residing in the mineral particles and grain boundaries. In addition, the forces related to the weight of overlying rock that are concentrated in the vicinity of a tunnel opening are transmitted through mineral grains and, depending on the physical properties of these grains, a variety of responses can be expected.

Several mechanisms operate to change the shapes and/or orientations of mineral grains under directed forces. Among these are the following:

(1) Microfracturing in brittle minerals without loss of cohesion.

(2) Microfracturing with loss of cohesion and more or less dispersion of the fragments. This process affects brittle minerals more so than soft or flexible minerals.

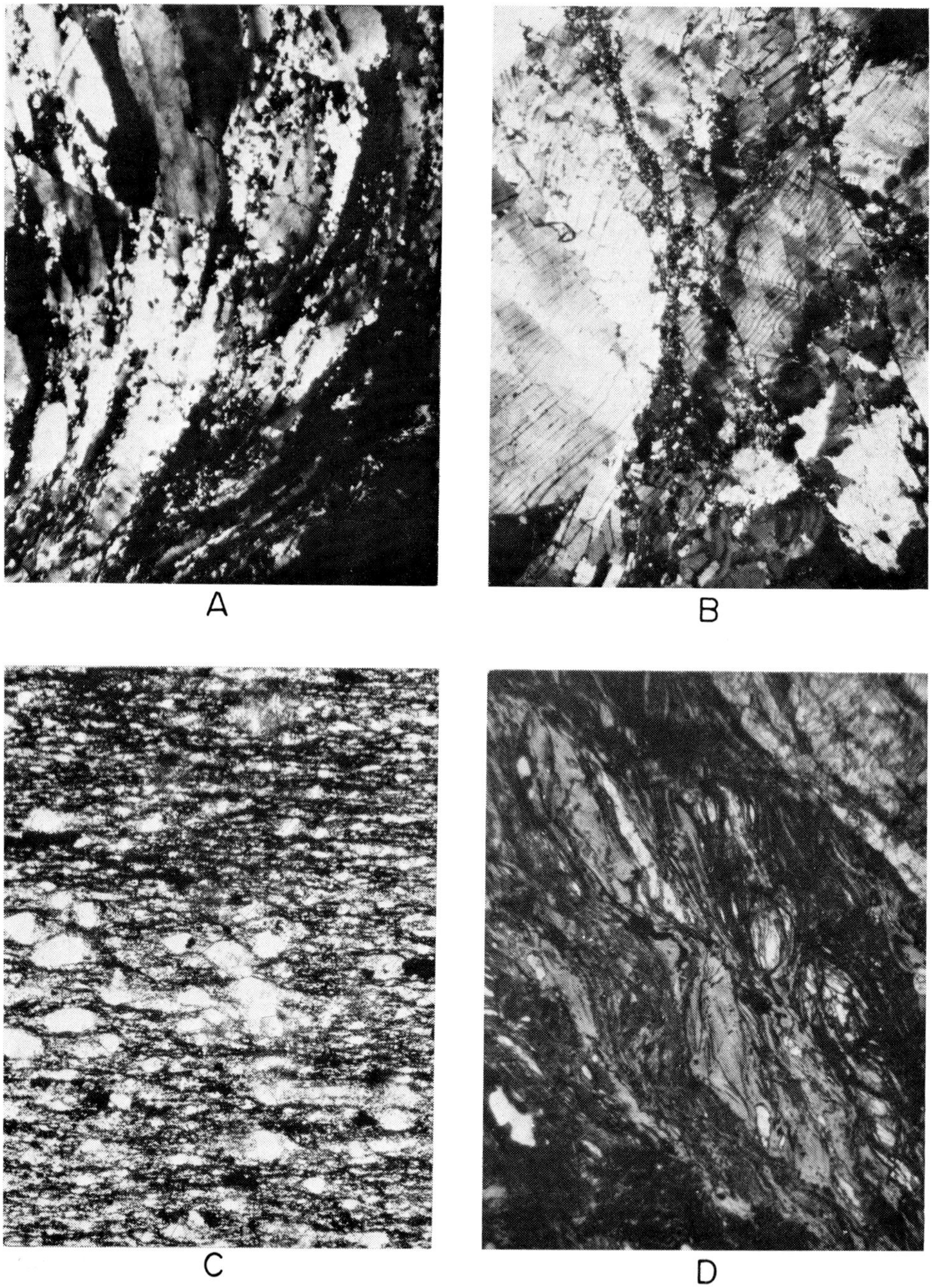

Fig. 33. Some modes of deformation of mineral grains in stressed rocks as seen in photomicrographs in polarized light. About 20×. A. Crushed and annealed quartz. Deformation in quartz produces uneven optical extinction. B. Plagioclase feldspar is fractured, crushed, bent, and twinned as a consequence of deformation. C. Quartz and feldspar grains (clear) in a sheared sedimentary rock have been rotated and flattened without loss of cohesion. D. Mica flakes have been bent and dragged out in a shear zone.

(3) Rotation involving, especially, platy, micaceous, linear, or tabular minerals.

(4) Gliding, which allows deformation by lateral dislocation of adjacent segments of a crystal by movement along internal crystal planes.

(5) Twinning, accompanied by physical dislocation of adjacent elements of crystal structure producing mechanical twins (as opposed to growth twins). Twinning is restricted to only one to three or four planar directions.

(6) Elastic strain, which results in bending of crystals without rupture, or, under hydrostatic pressure, in a volume decrease.

(7) Recrystallization, a mechanism which is promoted by pore water and results in changes in shape, size, or orientation of a mineral grain by simultaneous solution and redeposition at a localized site.

(8) Recrystallization resulting in the development of new minerals not originally present in a rock. This mechanism is promoted by the presence of pore water, and in a directed force field tends to create prismatic, platy, or micaceous minerals from preexisting minerals of different shapes.

The various mechanisms listed above represent responses of minerals to forces in a manner such as to tend to nullify the stresses associated with the forces. A net result is stored energy that is capable of partial release when the physico-chemical environment undergoes a notable change. Such energy should be expected to be concentrated in rocks, especially elastic rocks, that have been intensively and extensively deformed.

Fig. 33 illustrates several kinds of deformation in mineral grains in rocks that have been stressed.

DEFORMATION OF ELASTIC AND QUASI-ELASTIC ROCKS

Elastic rocks over short-time intervals and in the absence of time-dependent creep deform up to the point of rupture mainly by deformation within the mineral grains and along the contact zones of minerals. If any of the other mechanisms of mineral deformation are effective, either during short or very long time intervals, a rock body containing these minerals deforms elasticoviscously and assumes a new fabric, and, under some circumstances, a new mineralogy. Rupture, producing joints and faults, relieves some of the stresses in rocks and the minerals in them, and is an end product of deformation. Generally, changes in the shapes of large rock bodies must be considered as nonreversible because of the nonelastic behavior of such bodies over long terms. Additional changes in shape can come about on a large scale only by the superimposition of a new array of forces.

DEFORMATION OF PLASTIC ROCKS

Plastic rocks are weak rocks which deform easily without essential loss of cohesion. Most natural plastic materials are very fine-grained, typically containing abundant clay minerals and clay-sized particles of other minerals. Noncrystalline, amorphous or colloidal substances, and adsorbed or absorbed water also may be present in notable amounts. The very small size of mineral particles promotes development of surface forces of attraction, especially in the presence of pore water, that explain the ability of many plastic materials to be extensively deformed without rupturing. Fractures, when they develop in plastic materials, usually are a consequence of tension. Plastic rocks are characterized by differing degrees of stiffness, and a close correlation exists between stiffness and water content. Very stiff plastic rocks behave in a manner approaching that of elastic rocks and shear fractures as well as tension fractures may develop.

Deformation in plastic rock bodies, in terms of the mineral components, results mainly from sliding of grains past each other, and rotation of equidimensional and nonequidimensional grains without loss of cohesion of the rock mass. Generally, mechanisms causing changes in shapes of individual mineral grains are not operative.

Many plastic masses, partly through the loss of water and partly because of reorientation of mineral grains into stable configurations, gradually increase in strength with time, a process called *work-hardening*.

FOLDS IN ROCKS

Folds are most conspicuously developed in layered rocks, but originally isotropic rocks also may be folded, although the absence of layering makes detection of folds difficult. A comprehensive fold terminology is very complex, and a completely satisfactory classification is almost impossible to attain. A complete discussion of folds, with emphasis on folds in metamorphic rocks is presented by Turner and Weiss (1963). Other discussions of folds and folding may be found in a voluminous literature and in summaries in textbooks on structural geology, especially in a review by Whitten (1966).

The present treatment is not concerned so much with fold origin and terminology as it is with features of folded rocks that contribute to the behavior of rocks encountered in underground excavations, and, accordingly, only a few fundamental, elementary ideas will be discussed.

Two fundamental kinds of folds are recognized: *flexural-slip folds* (parallel folds) and *slip folds* (similar folds). Flexural-slip folds develop by slipping of adjacent layers past each other, and competent layers retain their original thicknesses. Slip folds (also called *shear folds*) are observed mainly in metamorphic rocks or sedimentary rocks deformed under considerable confining pressure, and folding results from slip (or shear) along closely spaced slip planes. Frequently such folds are accompanied by a conspicuous rock cleavage, and, for this reason they are sometimes also identified as *cleavage folds.*

Terms useful in describing the attitudes and directions of certain elements of folds are *hinge* (*hinge line*), *axial plane,* and *axis*. The hinge or hinge line passes through points of greatest curvature in a single folded surface and may be a straight or curved line. The axial plane (or surface) passes through the hinge lines on successive surfaces in a folded layered rock. The axial plane may be a planar or curved surface and may have any attitude from horizontal to vertical. The term axis has been used in the same sense as hinge line, a use that should be discontinued, except when the hinge is a straight line lying in the axial plane.

The geometry of gross dislocations in rocks producing folds ranges from simple to very complex, but even in a simple fold the distribution of strains and stresses, both on a megascopic scale and a microscopic scale, is not easily analyzed. In addition, the behavior of quasi-elastic rocks is quite different from the behavior of plastic rocks during folding. Adjustments during folding involve the total fabric of a

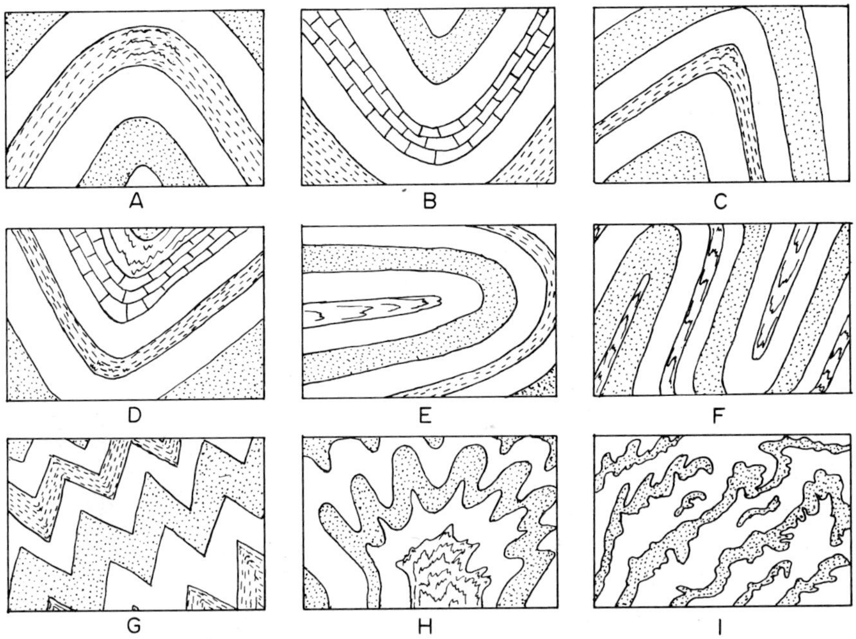

Fig. 34. Some shapes of folds in cross-section. No scale. A. Symmetrical anticline. B. Symmetrical syncline. C. Asymmetrical anticline. D. Asymmetrical syncline. E. Recumbent fold. F. Isoclinal folds with parallel limbs. G. Chevron, Kink, or accordion folds. H. Fan folds. I. "Flow" folds formed by plastic deformation, as in a metamorphic rock.

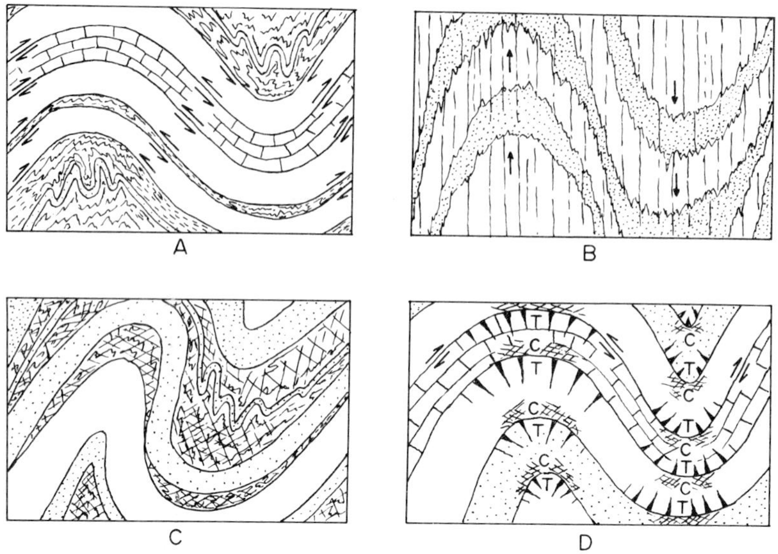

Fig. 35. Folds and subsidiary features developed during folding. A. Parallel (flexural-slip) folds showing relative movement of competent layers during folding. Plastic material (soft shale or mudstone) has been squeezed into the crest of an anticline and the trough of a syncline. B. Similar fold (slip fold) formed by shearing along a closely spaced set of parallel hairline fracture surfaces. A distinct rock cleavage exists in the direction of the fractures. C. Parallel folding in which a stiff semiplastic material has been squeezed into crests and troughs of folds. Semiplastic material is complexly folded and sheared. D. Sequence of parallel folds in brittle, competent rock, showing development of tension gashes or cracks (*T*) and complex shearing in zone of compression (*C*).

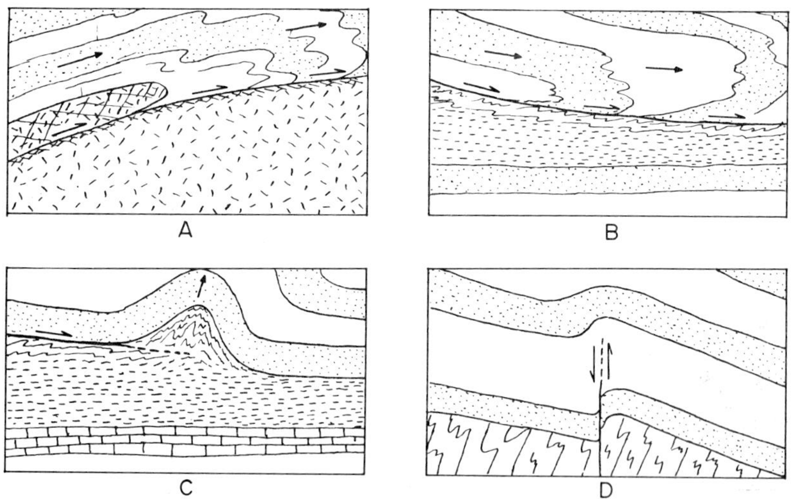

Fig. 36. Some folds associated with faults. Scale is hundreds to thousands of feet. A. Sedimentary rocks with a granite core are folded as they are displaced laterally over a surface of thrust faulting in granite. B. Sedimentary rocks have been displaced and folded by gravity-slip over a fault surface in shale. C. A fold formed by buckling as a mass of sedimentary rocks moves under gravitative forces over a surface of detachment in shale. D. A fault in basement rocks and in the immediately overlying sedimentary rocks dies out upward in a fold.

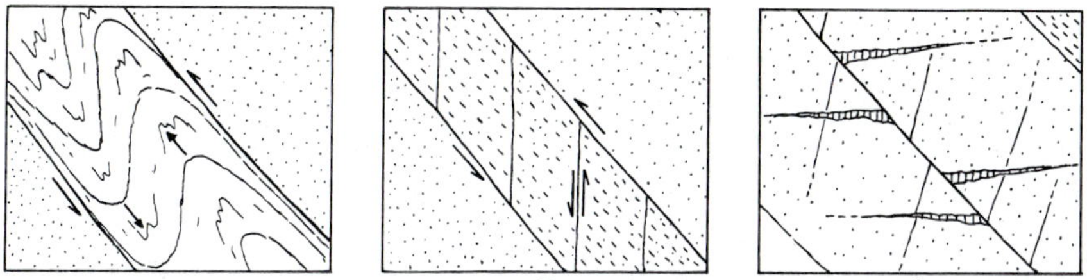

Fig. 37. Some small scale features associated with development of parallel folds. Bedding planes become fault surfaces by slippage of competent layers past each other. A. A layer of plastic material between competent layers is deformed into drag folds. B. A less competent layer is sheared by opposing movement in adjacent layers of more competent rocks. C. Open gash fractures produced by tension and tight shear fractures produced by compression develop during relative dislocation of two layers of competent rock.

rock, and include large and small scale dislocations, changes in shapes of mineral grains, shearing on a microscopic to megascopic scale, and development of tensional features.

Folding on a gross scale, and some of the subsidiary features associated with folding are shown in Fig. 34–37.

ENGINEERING ASPECTS OF FOLDED ROCKS

Folding of rocks, especially close folding, develops features that should be given careful consideration in planning and construction of tunnels. Aside from the fact that quasi-elastic rocks in folds store stresses in a highly anisotropic and complex manner, making it very difficult, if not impossible, to make an adequate stress analysis, there are other features that determine to a large extent the behavior of the rocks during excavation.

It should be expected that the number and variety of rock types encountered in a tunnel at right angles to the hinge line of folded stratified rocks will be greater than in a tunnel parallel or approximately parallel to the hinge line. Other situations can be visualized when the tunnel makes an oblique angle with the hinge line, or the hinge line is inclined.

In parallel folds in competent layered rocks, the movement of layers past each other is in fact fault movement, and stresses develop that produce tension and shear fractures. Tension and compression in the crests and troughs of folds produce cracks and shear intersections that control the flow of underground water, and many heavy water flows have been encountered in folds at or near the hinge-line portions.

The forceful movement during folding of plastic or stiff-plastic materials, such as shale, into the crests or troughs of folds, accompanied by plastic folding or complex, closely spaced shearing, commonly renders these materials completely incompetent, particularly when they are water saturated.

Folding of rocks during metamorphism at intermediate to high temperatures commonly is accompanied by moderate to extensive recrystallization and development of rock cleavage. Competency of originally plastic or semiplastic materials is improved by recrystallization of fine-grained mineral particles into coarser-grained, interlocking crystal aggregates. Except for the rock cleavage, metamorphic rocks in folded terranes are on the average much more competent than their unmetamorphosed sedimentary counterparts.

FAULTS

The mechanics of faulting are somewhat more comprehensible than the mechanics of folding. A fault is a consequence of accumulation of stresses in a rock to the point of rupture and displacement along the rupture. Faults are not always matters of grave concern in tunneling, and careful consideration should be given to their origin and physical nature before they are thought to be potential hazards in underground excavations.

Faults and associated joints are preferred channelways for circulation of groundwater, whether of surface origin or of deep-seated origin. Consequently, many faults are accompanied by extensive wall-rock alteration, either by surface waters or by hydrothermal solutions derived from a hot source at depth. Energy released by rupture and frictional heat generated by blocks of rock moving past each other cause notable increases in temperature along faults, sometimes sufficient locally to cause fusion of materials within the fault. If water is present in or near the fault, its temperature is elevated, and chemical reactions are promoted which may cause extensive alteration of wall rocks and materials within the fault. Chemical reactions generated in this manner simulate reactions produced by hydrothermal solutions derived from a hot deep source, and in zones of extensive fault dislocation tend extensively to modify the existing mineralogy. A common type of alteration is argillization, which produces clay minerals at the expense of feldspars, thereby introducing a high degree of plasticity into fault or fault-associated materials.

A simple classification of faults is based on the directions of relative movement of wall rock blocks. Accordingly, it is possible to identify *normal faults, reverse faults,* and *strike-slip faults.* In normal faults the block lying above an inclined fault surface has moved downward relative to the block under the fault. In reverse faults this movement is relatively upward. Strike-slip faults are characterized by essentially horizontal relative dislocation. Slip in either normal or reverse faults may be *down* or *up dip,* or *oblique.* The wall above an inclined fault surface is called the *hanging wall,* and the rock below the fault is in the *footwall* of the fault.

Many faults are the sites of repeated intermittent movements throughout a geologically long interval of time, especially in regions where the faults are associated with prolonged tectonic and igneous activity. Adjustments along the faults may be oscillating, and directions of relative movement of wall rocks may change radically as force fields causing movements change directions and intensities in response to progressive changes in the tectonic environment.

The magnitude of a fault generally is estimated by the width of the zone of fracturing associated with it and by the amount of total relative dislocation of wall rocks of the fault. It is true that faults that have caused major dislocations also commonly are accompanied by extensive zones of tensional and compressional fracturing. However, there are many fault zones hundreds of feet wide that have caused relatively small apparent offsets of adjacent rocks, presumably because movements along the fault were oscillatory, with numerous reversals of directions of relative movement. This observation suggests that it is unsafe to estimate the extent of subsidiary fracture development associated with a fault solely on the basis of measured net displacement. In oscillatory movements along faults the *total* displacement in all directions of movement might measure tens or hundreds of feet along a fault that shows a net dislocation of reference points of only a few feet. Under this circumstance it should be expected that subsidiary features such as drag folds and tension and shear fractures in the walls of the fault will be interrelated in a very complex manner.

Faulting as related to competency of rocks

Fault dislocations produce entirely different subsidiary features in high-strength, brittle rocks as compared with those in weak, plastic or semiplastic rocks. Examples of faults in rocks with differing degrees of competency are shown in Fig. 38.

Fault filling and gouge

Many faults contain sheared and crushed materials or have been filled with minerals precipitated from ground water solutions, either of surface origin or coming from depth as hydrothermal solutions (Fig. 39). Breccia filling of local origin is found in some faults and consists of angular to rounded brittle rock fragments derived from adjacent competent rocks and rolled and abraded by movements of adjacent walls of the fault. Commonly this breccia contains a matrix of finely ground materials, or in some instances, voids partly or completely filled with minerals deposited from solutions circulating through the breccia. Some breccia fillings in faults clearly are not derived from adjacent rocks and have been forcefully injected into the fault from a more or less distant site. Such breccias are called *intrusion breccias.* Many intrusion breccias prove to be rock debris forced upward

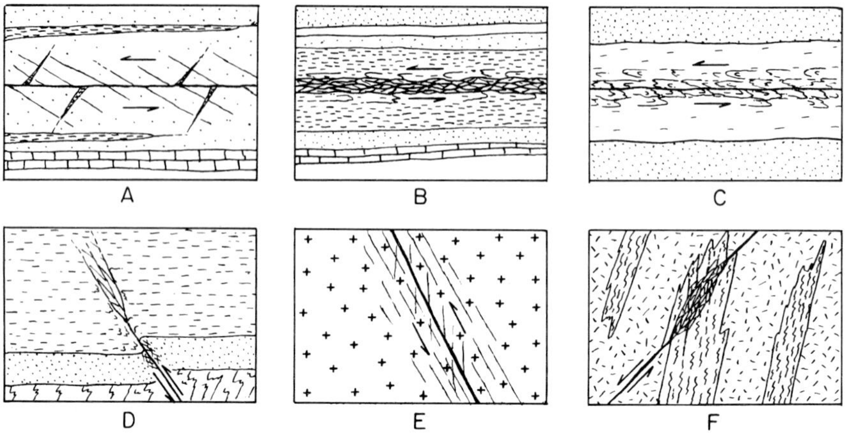

Fig. 38. Some examples of faults in materials of differing degrees of competency. A. Bedding plane fault in brittle rock develops associated shear fractures and tension (gash) fractures. B. Bedding plane fault in closely bedded semi-plastic shale develops closely-spaced, intersecting shears with a braided aspect. C. Bedding plane fault in poorly stratified semi-plastic material produces a wide zone of drag folds. D. Fault in competent, brittle rocks dies out in weak shale. E. Fault in a crystalline igneous rock develops subsidiary inclined shears and parallel sheeting. F. A fault in an igneous rock changes character in passing through a mica-rich metamorphic rock.

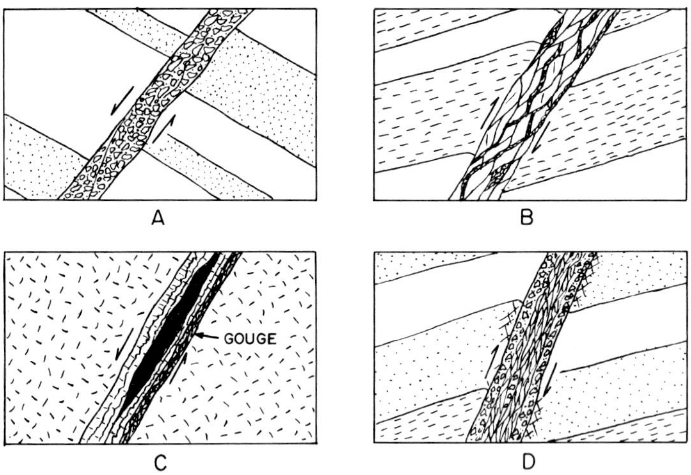

Fig. 39. Various kinds of fault fillings. A. Fault breccia. B. Braided fault with intersecting seams of gouge. C. Vein in an igneous rock with a gouge layer adjacent to foot wall. D. Gouge, which by movement concentrated in the center of the fault has rotated and forced larger angular fragments toward walls of fault.

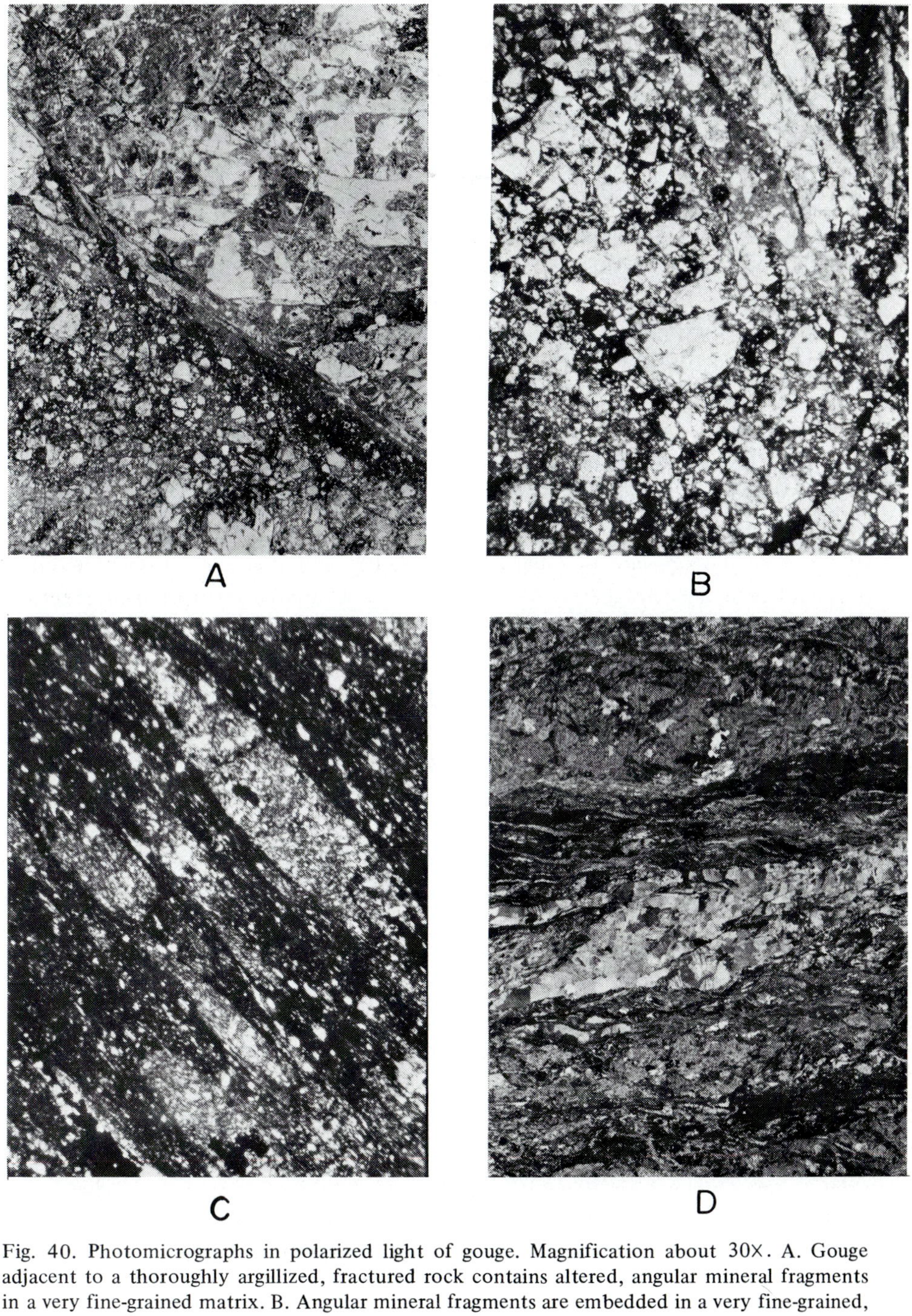

Fig. 40. Photomicrographs in polarized light of gouge. Magnification about 30×. A. Gouge adjacent to a thoroughly argillized, fractured rock contains altered, angular mineral fragments in a very fine-grained matrix. B. Angular mineral fragments are embedded in a very fine-grained, almost glassy matrix containing opaque iron oxides. C. Crushed mineral fragments have been dragged into lenticular segregations by shearing. D. Clayey gouge contains lenticular segregations of altered mica and argillized feldspars.

along a fault by an advancing igneous intrusion which pushes fragments of wall rocks and fractured, congealed igneous materials ahead of the intrusion.

An interesting phenomenon is injection of breccias downward into faults in brittle crystalline rocks. Basement rocks underlying sedimentary sequences commonly contain *breccia dikes* derived from the sedimentary rocks above by a mechanism that is not fully understood. In some examples, breccias of sedimentary rocks penetrate faults in granites and brittle metamorphic rocks for distances of hundreds or thousands of feet. Such injections probably occurred catastrophically, possibly impelled by high vapor pressures in water caused by frictional heat generated along faults. Similar breccia dikes also may be injected into overlying rocks, presumably by the same mechanism.

Gouge is thoroughly crushed and comminuted rock formed by grinding action between moving adjacent walls of a fault. Gouge formed in the presence of water commonly contains clay minerals in addition to clay-size particles of other rock minerals. The physical character of gouge as seen under the microscope is shown in the photomicrographs in Fig. 40. Gouge, whether in large faults or in minor subsidiary fractures, generally deforms plastically, and under pressure it may squeeze slowly into underground openings. Clay minerals and the small size of abundant micron-dimensioned fragments of other minerals promotes adsorption of water, so that, in addition to squeezing pressures, gouge may produce swelling pressures when it comes into contact with ground water or humid tunnel atmospheres. Rocks containing intersecting gouge layers are highly incompetent because gouge has little or no bonding strength.

Owing to its generally fine grain size, gouge is essentially impermeable to movement of underground water, and faults filled with gouge are poor aquifiers. Some relationships between faults and fault gouge are shown in Fig. 39. Gouge layers sometimes are accompanied by breccia. Movement within the gouge tends to rotate and force larger rock fragments out of the gouge into lenses within or at the sides of gouge layers.

Slickensides are scratches and grooves produced by movement of rock masses past each other in faults. They are observed not only on solid rock surfaces, but also in the soft material in gouge. Slickensides serve a limited use in determining the direction of relative dislocation of the wall rocks of a fault. Frequently, slickensides have many orientations, suggesting inconstant, multiple-direction, oscillatory dislocations along a fault.

Wall rock alteration associated with faults

Water of surface origin produces weathering alterations along faults and associated joints that may extend to depths of hundreds or thousands of feet in areas of considerable topographic relief where deep circulation is promoted. However, the most spectacular changes in wall rock composition and fabric are brought about by circulation along faults and associated joints of warm to hot water of deep-seated origin. Hydrothermal solutions (see Chapter 2) precipitate a wide variety of mineral deposits in faults to form veins, and spread out from faults along cracks to alter wall rocks. Alterations associated with some large faults extend tens or hundreds of feet into the wall rocks and, in general, tend greatly to reduce the competency of the rocks.

Except for certain types of alteration at high temperatures, such as tourmalinization or feldspathization, most kinds of hydrothermal alteration cause a dramatic decrease in the average grain size of the host rock and by conversion of larger mineral grains, especially feldspar grains, into aggregates of sericite or clay minerals thereby reduce the strength of a rock. Silicification, on the other hand, increases the competency of initially weak rocks, and fault zones accompanied by extensive deposition of silica within the faults and in the wall rocks frequently are more competent than the original rocks.

Hydrothermal solutions not only deposit materials but also dissolve materials.

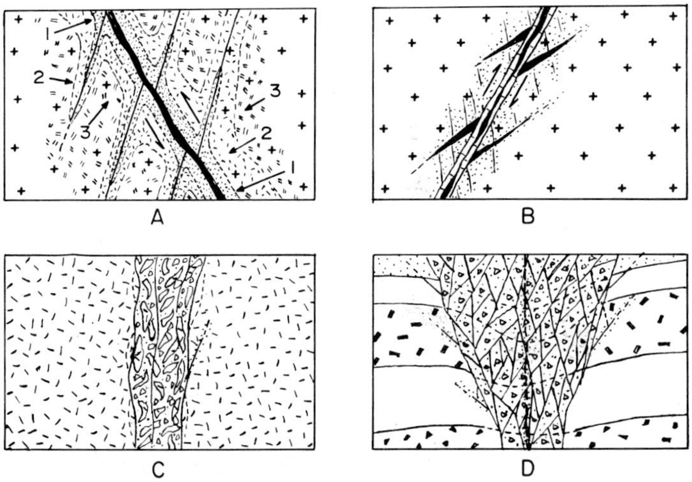

Fig. 41. Some effects of circulation of hydrothermal solutions along faults.
A. A metalliferous vein (black) is accompanied by zoned alteration in igneous wall rocks. A zone of silicification and sericitization (*1*) near the vein and subsidiary fractures grades outward into a zone of kaolinized feldspars, (*2*) and finally into a zone of alteration containing mixed-layer illite–montmorillonite (*3*).
B. Fault and adjacent gash fractures are filled with minerals deposited from solution. Argillic alteration permeates wall rocks (stippling).
C. A corrosive hydrothermal fluid enlarges opening along a fault by solution of the wall rocks. Walls of opening collapse into opening to form a *slab breccia.*
D. Corrosive hydrothermal solution dissolves wall rock along a fault and its subsidiary fractures to produce a *collapse breccia.* This phenomenon is sometimes described as *solution stoping.*

The chemical interplay between a hydrothermal solution and the wall rock of a fault is a complex process of exchange of materials. Some examples of features associated with circulation of hydrothermal solutions along faults are shown in Fig. 41.

Major faults

Mountain building and extensive dislocation of segments of the earth's crust have occurred throughout the geologic past and are active in some areas at the present time. Folding and faulting are a consequence of crustal deformation, and many mechanisms operate to produce a variety of large and small structural features. Some faults may be traced at the earth's surface for tens to hundreds of miles

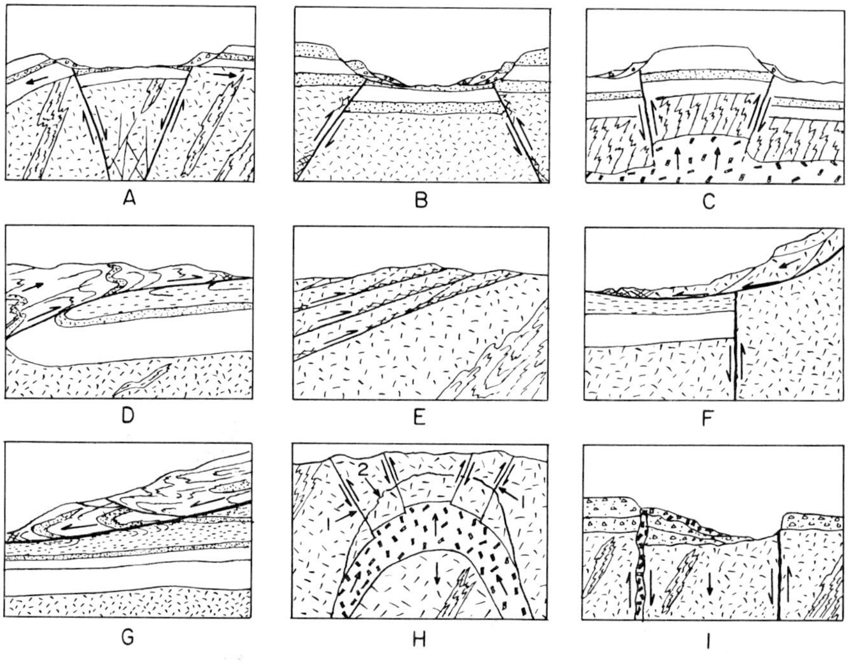

Fig. 42. Some mechanisms of formation of major faults. Scale is in miles or tens of miles. A. Tension failure and collapse of a wedge-shaped block along normal faults to form a graben. B. A rift valley is formed by dislocations along steep reverse faults. C. A deep-seated igneous body lifts a block of the earth crust along steeply dipping reverse faults. D. Sedimentary rocks are extensively deformed over a thrust fault. E. Flat plates (nappes) of crystalline rocks develop during thrust faulting. F. A gravity slip fault causes complex lateral movement of crystalline rocks over a curved shear surface. G. Sedimentary rocks are complexly deformed over a gravity-slip surface. H. An igneous intrusion produces shear fractures (1) and later a collapse zone of fracture (2). These fractures may be occupied by dikes. I. A caldera is formed by subsidence of a large block in a region of extensive surface volcanic activity.

and have greatly affected the rocks associated with them. Examples of major faults are the San Andreas fault in California, the Great Glen fault in Scotland, faults associated with the Rhine graben, and the great rift valleys of East Africa.

Because of the limited scope of the present treatment the origin and world distribution of major faults cannot be given extensive consideration. Instead, by means of the diagrams in Fig. 42, the nature and origin of some kinds of major faults are indicated. Fig. 42 does not include a representation of major strike-slip fault, sometimes called a *transcurrent, tear,* or *wrench fault.*

JOINTS

A *joint* is a rock fracture along which there has been little or no visible displacement in a direction parallel to the plane of the fracture. It is the absence of movement that distinguishes a joint from a fault, but incipient faults along which there has been very slight movement generally are classified as joints.

A *joint set* consists of several essentially parallel joints, whereas a *joint system* contains two or more intersecting joint sets. Regional geological studies often reveal a systematic geometrical relationship between joint systems and faults and folds, especially where the joints developed at the same time and in the same stress field. Examples of joints of systematic orientation formed by tension and compression associated with folding and faulting have been cited in earlier discussion in this chapter (see Fig. 35, 37 and 38).

Many situations prevail where joints are not systematically relatable to any larger structural features, and it is evident that joint sets of different orientations have formed at different times and from different causes.

In field investigations an attempt should be made to distinguish between surficial joints and joints formed at depth. Surficial joints develop by glacial loading and unloading, slump in oversteepened slopes, contraction and expansion of rocks exposed to the atmosphere, unloading of rocks as a consequence of stream erosion, and the like. Such joints do not penetrate rocks below the surface for any great depth, and in general are of no consequence in the excavation of tunnels except in the portal areas or where surface cover in a tunnel is shallow.

Field study and graphical representation of multiple joints

Joints are mapped in the field during routine geologic investigations or noted during inspection of aerial photographs. In field investigations the strike, dip, and spacing of the joints are determined, and an effort is made to ascertain whether the joints have formed by tension or compression and if alteration has been localized along the joints. If several sets of joints exist in an intersecting pattern, age relationships of the various sets should be interpreted, if possible.

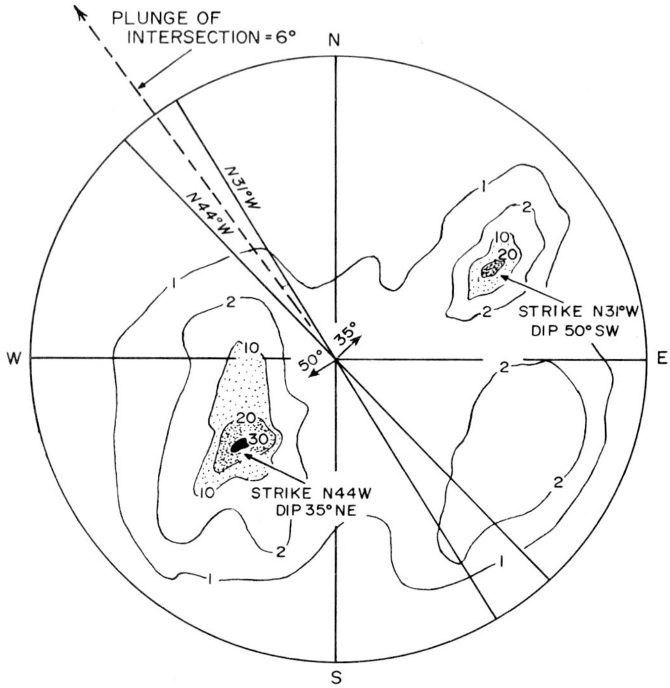

Fig. 43. Contoured equal-area plot of a joint system containing two joint sets. Lower hemisphere.

A typical study of joints yields large numbers of measurements. When strikes and dips have been measured, a statistical study by means of equal-area plots serves a useful purpose. Construction of equal-area plots is summarized in Chapter 7.

Fig. 43 is a contoured equal-area plot of a large number of joint measurements. Two concentrations of poles of joint planes indicate the existence of two sets of joints with intersections plunging at a small angle from the horizontal. The equal-area plot permits prediction of the attitudes of joints as they will be encountered in a tunneling operation of specified direction, and permits an assessment of the caving tendency and excavation overbreak.

Another effective device for graphically representing the distribution and attitudes of large numbers of joints (or faults) is the *joint rose,* also called a *star diagram.* The joint rose is most useful when the directions of large numbers of joints have been measured, but the dips are not known. Such data typically are assembled from aerial photographs where no opportunity has been provided to measure dips in the field.

Fig. 44 is a rose diagram of two sets of joints with strikes nearly at right angles. In preparing the plot the total number of joints counted in each 10-degree sector

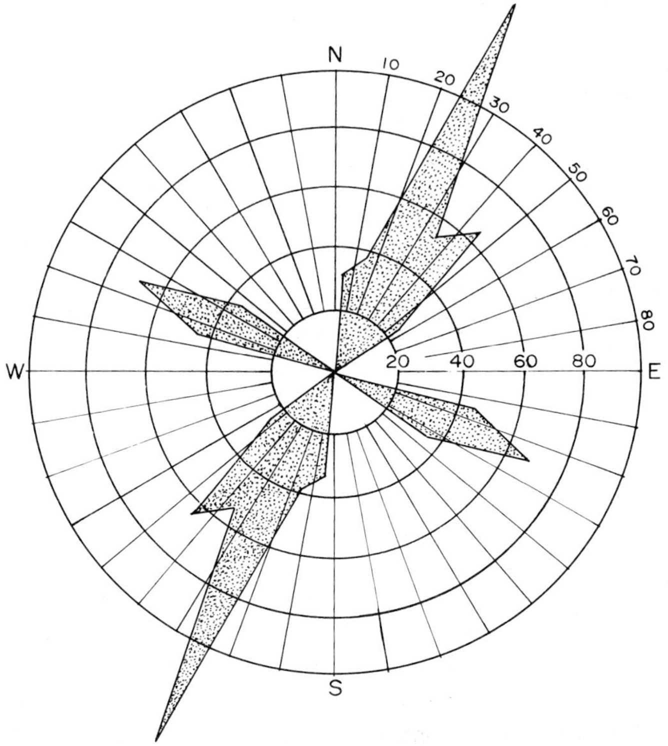

Fig. 44. Joint rose showing number of joints counted in each 10-degree sector. The plot shows two sets of joints with average strikes about N25°E and N65°W.

was plotted on a radial line bisecting the sector and lines were drawn connecting the points in the various sectors.

Spacing of joints and characteristics of joint surfaces

The behavior of jointed rock in underground excavation depends on several factors, but especially on the spacing, attitudes, and surface characteristics of the joints. Clearly the spacing and attitude of the joints determines the number of joints that will be encountered in the tunnel cross-section, and, depending on the size of the opening, may make the difference between competent or incompetent behavior of the rocks. Consequently, every effort should be made prior to tunnel construction to estimate the spacing and attitude of joints by examination of surface rock exposures, drill cores, and by use of applicable geophysical methods.

The attitudes of joints in tunnel excavations, as well as the spacing, determines the ease with which joint blocks are detached from the arch and walls of a tunnel under the influence of gravity or as a consequence of blast damage. However, there

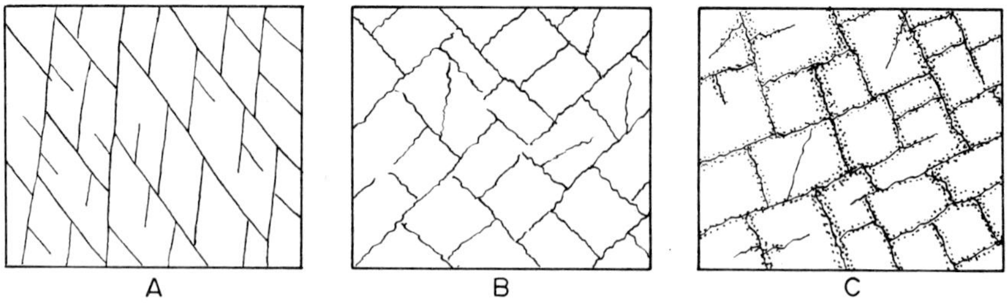

Fig. 45. Various aspects of joints. A. Intersecting smooth joints. B. Joints with rough surfaces in blocky ground. C. Joints with rough surfaces, but wall rocks of joints have been altered to clay minerals (stippled).

is an important difference in the behavior of joint blocks, depending on whether joint surfaces are smooth or rough, and on whether the joints are altered or contain clay minerals, and/or water.

Joint surfaces that are smooth (Fig. 45A) commonly display slickensides and contain crushed materials that resulted from slight dislocations during mass adjustments within the body of rock containing the joints. Closely spaced intersecting joints of this type frequently create highly incompetent, caving rock in tunnels. Joints with rough surfaces tend to develop in crystalline rocks by tensional stresses (Fig. 45B) and are not easily disengaged from rock exposures because of the high degree of frictional resistance along the joint surfaces. Surface solutions or ground waters of deep-seated origin find natural channelways of circulation along joints, and many joints, whether smooth or rough, have localized argillic alteration in the walls of the joints, greatly increasing the ability of joint blocks to slip past each other (Fig. 45C).

Original joints in igneous rocks

Joints in igneous rocks have a variety of origins, but there are several kinds of joints that can be identified with the emplacement and cooling of igneous bodies. Some joints develop in the congealed outer portions of igneous intrusions as a consequence of intrusive pressure from still liquid magma below. However, the most abundant and conspicuous joints originate from cooling contraction. Shapes assumed by joint blocks range from prismatic, to blocky-equidimensional, to platy, depending on the shape of an igneous body and its cooling history. Prismatic or platy shapes are generally associated with quickly congealed tabular igneous bodies such as dikes, sills, and lava flows. Blocky shapes are common in larger igneous bodies, where cooling was slow. Fig. 46 illustrates two examples of original joints in igneous rocks.

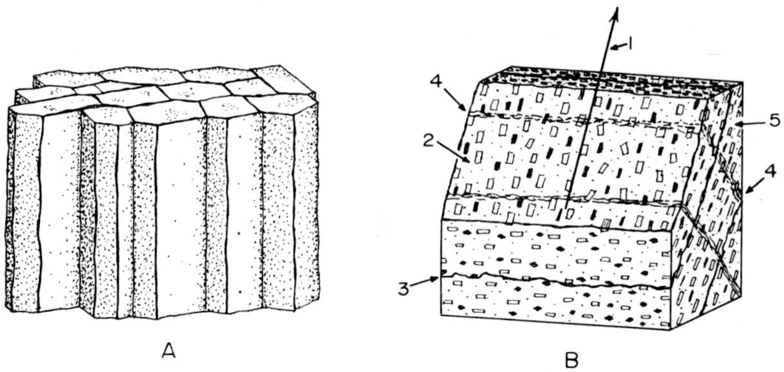

Fig. 46. Original joints in igneous rocks. A. Columnar joints producing prismatic shapes in fine-grained-tabular igneous body. B. Block diagram showing flow structures in a crystalline igneous rock. *1* = linear flow structure parallel to prismatic minerals; *2* = platy flow structure defined by tabular minerals in parallel arrangement; *3* = contraction joint parallel to platy structure; *4* = tension joints formed at right angles to flow lineation; *5* = possible joint direction parallel to flow lineation.

Original joints in sedimentary rocks

Joints developed by faulting and folding in sedimentary rocks have been described in an earlier section of this chapter. Other kinds of joints form during the burial and compaction of sediments, and include desiccation joints and joints from unequal compaction. Many brittle sedimentary rocks, such as limestones and hard sandstones, contain another kind of jointing that cannot be explained easily, yet it is almost universally present in such rocks whether they have been tectonically deformed or not. Fig. 47 shows joints of this type in undeformed sediments. Gener-

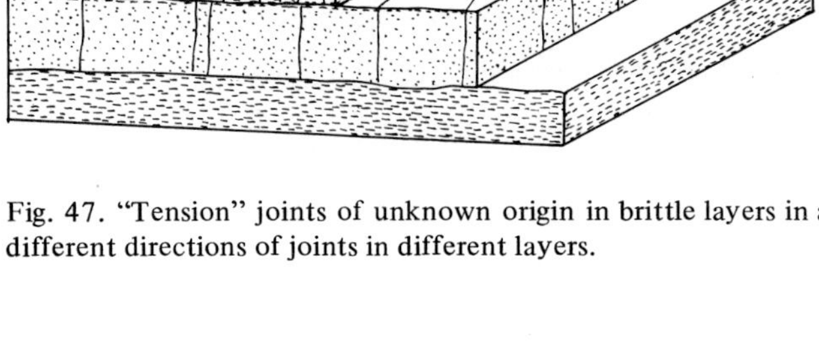

Fig. 47. "Tension" joints of unknown origin in brittle layers in a sedimentary succession. Note different directions of joints in different layers.

ally, the joints are nearly perpendicular to the bedding, but sets have different strikes in different layers. The joints appear to have developed by tension and usually have rough surfaces. An abundance of these joints in a sedimentary layer greatly increases the permeability of the layer, and it is not uncommon to encounter underground water flows issuing from one or several of these joints.

Original joints in metamorphic rocks

Original joints develop in competent layers undergoing metamorphism. However, in medium to high-grade metamorphism recrystallization tends to heal fractures that develop during deformation. In such rocks original dislocations are preserved in rock cleavage, and the rocks do not contain actual joints, although subsequent deformation frequently causes joints to develop parallel to the rock cleavage. Original joints are preserved in some low- to medium-grade rocks such as slates and rocks containing highly competent layers alternating with very weak layers.

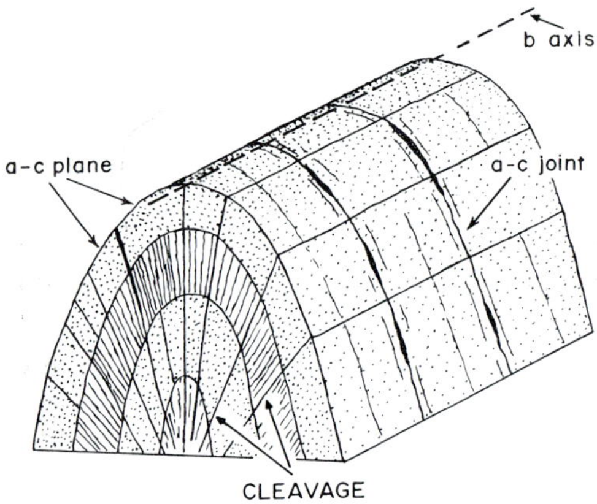

Fig. 48. A fold in competent rocks deformed by flexural slip during metamorphism. Original joints develop in competent layers parallel to the fold axis (*b*) and at right angles to the fold axes.

Fractures in highly competent layers of metamorphic rocks deformed by flexural-slip folding are shown in Fig. 48, together with conventional designations. Joints in highly competent layers in which there has been only slight rotation or recrystallization of minerals into the direction of the joints produce so-called *fracture cleavage.* Joints in layered metamorphic rocks are shown in Fig. 49.

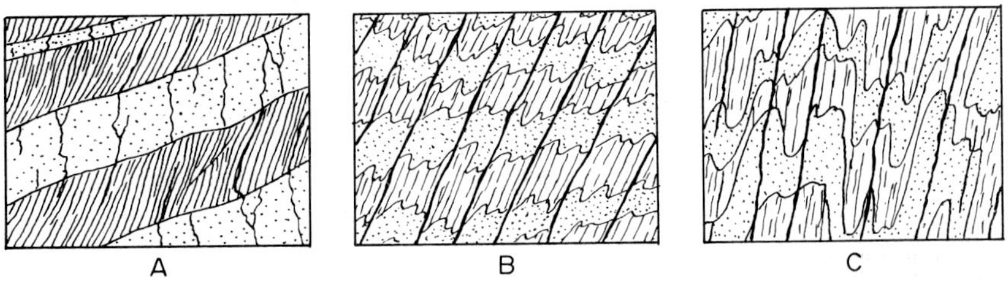

Fig. 49. Joints in layered metamorphic rocks. A. Joints occur in quartzite layers alternating with slate which displays slaty cleavage. B. Joints associated with kink bands in a layered rock. Joints are actually faults with slight displacement. C. Joints of late origin following rock cleavage planes in a highly deformed metamorphic rock.

TUNNEL SECTIONS IN FRACTURED ROCKS

Sections of tunnels in closely jointed and faulted rocks tend to be characterized by considerable overbreak when excavated by drilling and blasting. The shape of the tunnel cross-section reflects the spacing and attitudes of the fractures, especially the joints. Smooth joint surfaces or alteration along the joints notably increase the amount of overbreak. Widely spaced joints with rough surfaces and no alteration scarcely reduce the overall competency of wall rocks. Examples of overbreak in closely jointed, altered rocks are shown in Fig. 50 and 51, based on measurements in the Roberts Tunnel in Colorado. Fig. 50 shows average overbreak in a portion of the tunnel in which the joints dip steeply into the tunnel opening. The average

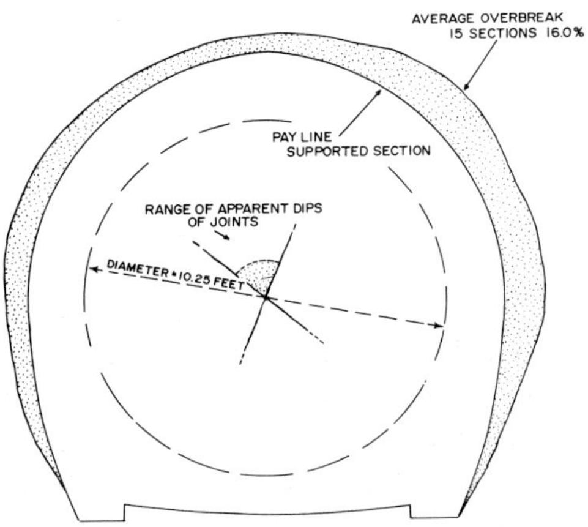

Fig. 50. Overbreak in closely jointed rocks in granite (quartz monzonite) in the Roberts Tunnel, Colorado. Diameter of finished, lined tunnel is 10.25 ft.

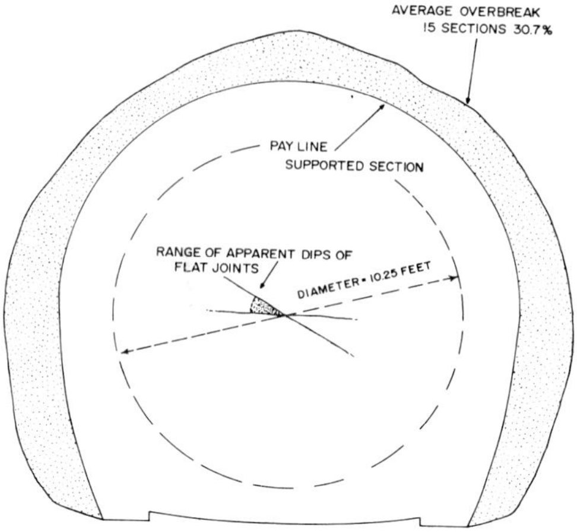

Fig. 51. Average overbreak in a section of the Robert Tunnel, Colorado, characterized by closely spaced, approximately horizontal joints. Note conspicious arching in rocks above pay-line. Joints localize argillic alteration.

spacing of the joints is about 2 ft. Note the elliptical overbreak section with the long axis of the ellipse approximately normal to the average apparent dip of the joints as they are observed in the tunnel heading.

Fig. 51 shows the average overbreak section in a portion of the tunnel characterized by closely spaced flat joints in the arch portion of the tunnel. Overbreak is extensive, and the tunnel cross-section has developed a conspicuous arch, a feature that also is noted in tunnels in flat-lying, closely bedded sedimentary rocks such as shales.

THE ARCHING TENDENCY

Tunnels in incompetent rocks tend to develop stable cross-sectional configurations with time. Except under the most extreme conditions of swelling, squeezing, or water-soaked running ground, it can be expected that highly-fractured, altered rocks will, within a period of a few days after first exposure in a tunnel opening, tend to stabilize, particularly if supports are installed so as to minimize caving. This observation is particularly pertinent in estimating the amount of support that is needed to prevent long-term movement of fractured and altered materials into the tunnel opening. Stabilization is a complex process involving removal of water, accompanied by stiffening of altered fine-grained materials and the building of a stable arch configuration by the blocks or fragments of relatively competent materials in the fractured ground.

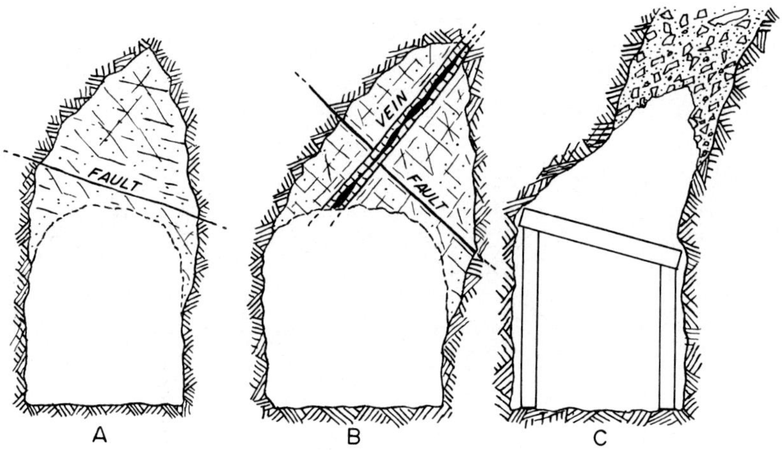

Fig. 52. Arching in fractured, altered rocks, producing stable tunnel cross-sections. A. Arching is related to a flat fault accompanied by closely spaced joints. B. Arch location is determined by a vein accompanied by joints and argillic wall rock alteration. C. Fragmented rock (muck) in a stope has developed a stable arch profile.

The writer has made many observations in abandoned mine workings, some a hundred years old, where timbers have rotted away and no longer provide the support deemed necessary at the time that the tunnels were driven. In spite of the fact that many of these tunnels follow highly-fractured and altered zones, the tunnels remain open because of the formation of stable arch profiles. In addition, completely fragmented waste rocks (muck) in old stope fillings commonly stabilize with time and remain in the stopes without support.

Some examples of the arching tendency in fractured rocks are illustrated in Fig. 52 and demonstrate the tendency for fractured rocks to establish stable cross-sections. Consideration of the ability of fractured rock masses to develop arches suggests that the maximum weight of caving ground that should be supported in the arch of a tunnel does not exceed the dead load weight of a rock mass with a cross-section of the tunnel opening.

The arching tendency in underground openings is present in all kinds of natural materials, whether they are fractured or not. Rocks of low strength, especially layered rocks with bedding-plane discontinuities, may fail under stresses in and around the tunnel crown and, if unconfined, more or less rapidly develop an arch. An example of the initial stage of development of an arch is shown in Fig. 53.

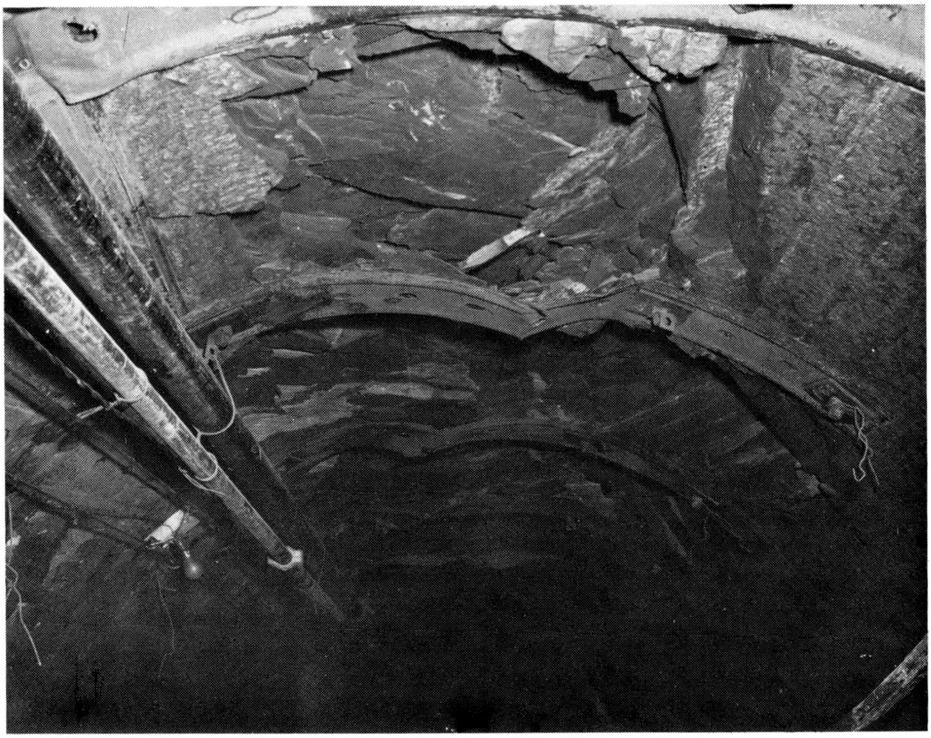

Fig. 53. Arching tendency results in failure of bedded shale by stress release in the crown of a mole-bored tunnel. Note deformation of light steel supports pinned with rock bolts. Oso Tunnel, San Juan-Chama project, Colorado-New Mexico. (Photo by U.S. Bureau of Reclamation.)

REFERENCES

Turner, F.J. and Weiss, L.E., 1963. *Structural Analysis of Metamorphic Tectonites.* McGraw-Hill, New York, N.Y., 545 pp.

Whitten, E.H.T., 1966. *Structural Geology of Folded Rocks.* Rand McNally, Chicago, Ill., 663 pp. plus index.

Chapter 6

HYDROGEOLOGY, TEMPERATURES, AND GASES IN TUNNELS

INTRODUCTION

Ground water, if present in large volumes and capable of free movement, presents a major hazard in tunneling. Estimation of the amounts of water and the location of heavy ground-water flows that might be encountered during a tunneling operation is very difficult, especially in areas of considerable geological complexity. In addition penetration by ground water of many kinds of altered, argillaceous, or closely fractured rocks promotes swelling, disintegration, or lubrication that produces capabilities in these rocks of slow to very rapid movement into a tunnel opening, greatly increasing the difficulty of the tunneling operation. Some of the most disastrous experiences in underground openings have been the result of intersection of heavy flows of water or highly fractured and thoroughly altered water-soaked rocks. It is essential that in planning for a tunnel every attempt should be made to identify the location and size of bodies of underground water whether concentrated in fissures or dispersed through masses of porous and permeable rock.

Warm or hot waters, with temperatures above 30–35°C (86–95°F) present special problems, especially in creating very uncomfortable to almost intolerable working conditions. Accordingly, efforts should be made during preliminary geologic investigations to identify conditions that might yield flows of warm to hot water into a tunnel.

Squeezing, swelling, or running ground in tunnels almost invariably is associated with ground water, either freeflowing water, pore water, or atmospheric moisture. Accordingly, because of this association, the characteristics of the materials that create tunnel hazards by moving into tunnel openings by expansion, squeezing, or free flow are reviewed in this chapter.

Gases encountered underground also may present problems, and frequently move through rocks along the same courses followed by flowing underground water, and rocks that are permeable to water also are permeable to underground gases.

WATER IN ROCKS

Water is present in rocks in a variety of states and associations with mineral grains and open spaces. A classification of the kinds of water in rock follows:

(1) Chemically bound hydroxyl (OH) in minerals as, for example, in mica.

(2) Chemically bound water (H_2O) in minerals, as in zeolites or hydrous sulfates.

(3) Liquid inclusions in minerals.

(4) Hygroscopic or absorbed water which is loosely held within a substance as in montmorillonite.

(5)Adsorbed or pellicular water which adheres to grain surfaces because of surface forces.

(6) Capillary water which occupies capillary openings because of surface attraction and surface tension.

(7) Free water or water of percolation capable of moving through rocks under a hydrostatic head.

Water in the first three categories is of no concern in the present context, but the other kinds affect the properties and behavior of rocks in underground openings. Another kind of water that is available for absorption or adsorption by rocks is atmospheric water which, in highly-humid tunnel air, is capable of causing changes in rock bodies, especially in those rich in clay minerals, adjacent to the tunnel.

Hygroscopic or absorbed water plays a particularly important role when it is associated with expanding clay minerals such as montmorillonite or mixed-layer illite-montmorillonite. A familiar example of a rock with absorbed water is bentonite, an altered volcanic ash rich in montmorillonite. The water content in bentonite is reduced by pressures of burial in a layered sequence, but exposure of the material to ground water or atmospheric moisture promotes expansion and swelling to many times the original volume. Similarly, rocks containing montmorillonite formed by hydrothermal or weathering alteration swell and disintegrate on contact with water or atmospheric moisture.

Adsorbed water, clinging to the surfaces of particles, is of slight consequence in coarse-grained or nonporous rocks, but is a matter for concern in even slightly permeable very fine-grained materials such as shales, mudstones, and extensively argillized rocks. Water molecules are polar, that is, they are characterized by positive and negative electrical charges on opposite sides of the molecules, a feature which explains the high surface tension of water. Surfaces of crystals and fragments of minerals commonly exhibit unsatisfied electrical charges, forces which attract polar water molecules and hold them with great tenacity in thin films. The surface forces that attract water in a loose aggregate of mineral grains are proportional to the total area of the surfaces bounding the mineral particles in the aggregate, and in fine-grained aggregates that are not highly lithified and/or pervasively cemented large amounts of water can be retained by adsorption. For example, if a cube with a surface area of six square centimeters is subdivided into cubes of the size of clay particles the total surface area will increase to about 31,400 cm^2. This relationship is evident in a comparison of relative amounts of adsorbed water in a typical sand, about 0.4 wt.%, and an unconsolidated clay, about 17 wt.%. Poorly consolidated, fine-grained rocks of low strength adsorb water on exposed surfaces in tunnel openings often causing so-called *moisture-slaking* and/or *adsorption swelling.*

Capillary water is held in rocks in openings of capillary size and is not found in crystalline rocks or well cemented clastic sedimentary rocks in appreciable amounts, but may be present in considerable amounts in poorly cemented clastic rocks.

Free water capable of moving through rocks either along cracks or open fissures or along interconnecting pore spaces in fragmental rocks is the cause of most major problems associated with water in tunneling operations.

POROSITY AND PERMEABILITY

Porosity is a measure of the volume of voids in a rock capable of containing water. A distinction is made between *primary* or *original porosity,* which is determined by the natural processes that resulted in the initial formation of a rock, and *secondary porosity,* which is associated with cracks and solution openings developed at a later time. Primary porosity is minimal in crystalline rocks, and, on the average, is considerably less than 1%. Maximum porosity exists in unconsolidated sedimentary accumulations and in fine-grained or glassy igneous rocks such as pumice or scoria filled with gas cavities. Some original porosities of consolidated and unconsolidated sediments are given in Table XIV. A distinction is sometimes made between *total porosity* and *effective porosity.* Total porosity is the per cent volume of all voids, large or small, and effective porosity is an expression of either the

TABLE XIV

Porosities of sedimentary rocks (vol.%)

Sediments	*Average*	*Extremes*
Consolidated:		
Sandstones	15	0.0–38
Limestone, dolomite, chalk and marble	4	0.1–43
Shale	20	9 –37
Unconsolidated:		
Loess	61	
Fine sand	46	
Very fine sand	48	
Sand–silt–clay	74	
Mud	88	
Silt	55	

volume per cent of voids accessible to filling by water or the volume of water that can be recovered in a porosity test, which, of course, does not include retained water of adsorption or capillary water. For example, a volcanic rock filled with isolated gas bubble voids may have a very high total porosity and a very low effective porosity.

Permeability measures the ease of transmission of water under a head and is measured with permeability apparatus (a permeameter). Permeability depends on the presence of available channels of flow, and, as for porosity, a distinction can be made between *primary or original permeability.* which is identified with a material in its original state, and *secondary or induced permeability,* which is associated with cracks or solution-formed openings. Permeability usually is expressed in terms of a *coefficient of permeability,* which is an element of Darcy's equation for flow of fluids through permeable media. Darcy's equation, simply expressed, is:

$$Q = KAH/L \tag{5-1}$$

where Q is the quantity of water flowing through a cross-section of area A, of a material having a coefficient of permeability K and for which the hydraulic gradient is H/L. Hydraulic gradient is defined by the ratio of head loss by frictional resistance, H, to distance L, in the direction of flow, or in more general terms, as $\mathrm{d}H/\mathrm{d}L$. Darcy's law is concerned with *movement* of water through permeable materials, and it does *not* state that flow takes place from low to high pressure, except when flow is horizontal. Instead, the term H, the head loss, is a measure of the difference between the piezometric pressures at two points at distance L apart, the decrease in pressure in a particular direction depending strictly on frictional resistance to flow.

In ground water hydrologic studies the coefficient of permeability commonly is identified by means of the *Meinzer unit,* expressed in terms of gallons per day per square foot of cross-section (gpd/ft.^{2}). In eq. 5-1, units used in accord with the definition of the Meinzer unit are gallons per day at 60°F for Q, square feet for A, feet of water for H, and feet for L. In Europe the coefficient of permeability is expressed in metric units. Another unit in use is the *Darcy,* which measures the flow in cubic centimeters per second of a fluid of unit viscosity through a cross-section of one centimeter under a pressure gradient of one atmosphere per centimeter.

Notable primary permeabilities (more than about 1–10 gpd/ft.^{2}) are present in relatively few rocks. Permeability in crystalline igneous, metamorphic rocks and fine-grained sedimentary rocks such as shales and mudstones is at a minimum, and the permeability coefficient is less than 0.01 gpd/ft.^{2}. Maximum primary permeability is found mainly in clean sandstones with little matrix and/or cement, as in petroleum reservoirs and in some notable sandstone aquifers such as the Dakota sandstone in the western United States. Inflows of water into tunnels from solid rock aquifers with only primary permeability rarely presents difficulties in tunneling. Water flows from faults and joints intersecting such rocks are an entirely different matter.

It should be noted that in tunneling in unconsolidated surficial deposits, such as stream deposits and glacial deposits, the probability exists that highly-permeable deposits may be encountered with permeabilities of the order of 10–20 gpd/ft.^{2}. In solid rocks such high permeabilities result only from the local development of secondary permeability by fracturing or solution channelways.

Tests of primary permeability are generally made in cores obtained during exploratory drilling of tunnel sites.

PERMEABILITY IN VOLCANIC ROCKS

Maximum permeability is encountered in some lava flows, especially those of basaltic composition. Basaltic lava flowing over the earth's surface frequently produces thick, loose accumulations of blocks resulting from the break up of chilled surfaces while the still liquid interior of flows continue movement. In addition, flow in the interior of partly congealed flows sometimes leaves lava caves. Subsequent flows or explosive pyroclastic deposits cover the blocky surfaces of older flows and preserve zones of very-high permeability which allow easy movement of large volumes of underground water. Silicic volcanic materials generally have very-high viscosity, even at temperatures of complete liquidity, and do not form lava flows of any significant lateral extent. Instead, most layered deposits of volcanic rocks of silicic composition are pyroclastic, and, ordinarily, are not characterized by notable permeability, except in aggregates of explosion breccia containing only small amounts of finer grained matrix.

WATER TABLE AND AQUIFERS

Rock aquifers are bodies of rock that permit relatively easy flow of water through them because of notable primary and/or secondary permeability. Impermeable rock bodies enclosing aquifers are called *aquicludes.* Water percolating downward through rocks with laterally and vertically widespread permeability saturates the rocks below a level called the *ground-water table.* Water moving downward along interconnecting fractures in impermeable rocks reaches a level below which the fractures are filled, that is called the *water level,* or the *peizometric level.* Fig. 54 illustrates some of the terms commonly used to describe underground hydrologic relationships.

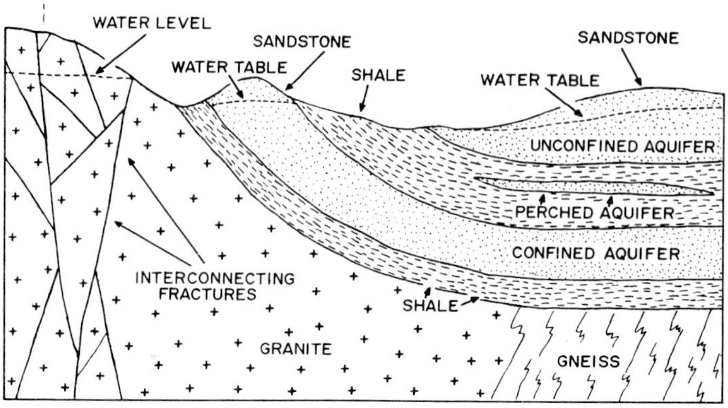

Fig. 54. Diagram showing use of several hydrologic terms. Permeable sandstones alternate with impermeable shales in a sedimentary sequence resting on crystalline rocks.

SOURCES AND TEMPERATURES OF UNDERGROUND WATER

By far the greater quantity of water in rocks in the upper portion of the lithosphere has been derived from the atmosphere and the hydrosphere at the earth's surface. Such water is described as *meteoric water,* as opposed to *juvenile water,* which has its source in the hot, deeper parts of the earth. Some of the water in deep aquifers or perched aquifers, so-called *fossil water,* comes from adjacent water-bearing sedimentary rocks, such as shales, which lose water as they are compacted under the weight of overlying rocks.

Large volumes of steam and hot water accompany volcanic activity in many areas, and although much of this water probably is ultimately of meteoric origin, there is little question that at least some of the water originally was contained in molten rocks at depth and escaped as a consequence of cooling and crystallization at or near the earth's surface or at depth.

Metamorphism at moderate to great depths tends to drive water out of rocks, providing large volumes over very long periods of time.

Underground temperatures of water, rock, and humid tunnel atmosphere that exceed about 32°C or 90°F present very difficult working conditions, usually requiring extensive refrigeration and/or rapid circulation of tunnel air. Consider the difficulties created in the Simplon Tunnel in Europe when at a depth of 7,000 ft. water was encountered with a temperature of 133°F (56°C).

In planning underground excavations every consideration should be given to the possibility of intersection of rocks and water flows with temperatures above tolerable limits, especially in regions of complex geology that has resulted from extensive crustal dislocation and igneous activity within the last tens of millions of years.

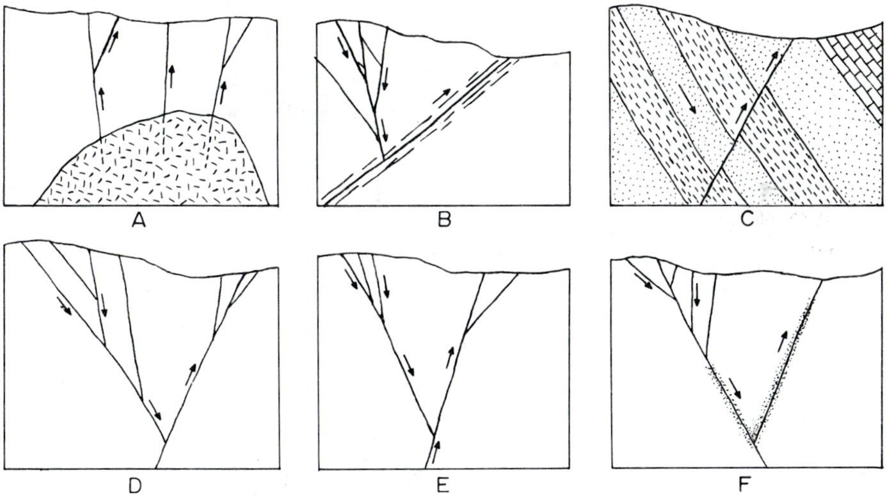

Fig. 55. Hypothetical geological conditions leading to development of warm to hot water below earth's surface; A. Hot water escapes from a still hot, crystallized body of igneous rock at depth. B. Water circulating along a fault is heated by frictional heat developed by dislocation of rocks along the fault. C. A sandstone aquifer conducts water to considerable depth where its temperature is increased because of the geothermal gradient. D. Same as C, but circulation of water is controlled by fractures. E. Water moving downward mixes with hot water from an unknown, deep source. F. Circulating ground water is heated by exothermal chemical reaction with wall rocks (stippling).

Sources of heat in underground water and associated rocks are of the following kinds:

(1) Heat slowly conducted from the hot interior of the earth. This heat is of little concern in the present context.

(2) Residual heat in congealed igneous rocks and adjacent heated country rocks or heat from molten rocks at depth.

(3) Heat generated by chemical processes such as reaction between ground water and country rock.

(4) Heat introduced into rocks by igneous liquids or gases from deep hot sources.

(5) Radioactive heat from decay of radioactive elements.

(6) Heat generated by friction during fault or fold dislocation of rocks.

Some hypothetical geological conditions producing warm to hot underground waters are shown in Fig. 55.

The presence of active thermal springs in an area provides a warning that hot water may be encountered at depth, but topographic and geologic conditions may be such that issuance of thermal water at the earth's surface cannot be realized. However, if ground water reaching the earth's surface in springs is only a few degrees warmer than would be expected for circulation of surface-derived water at shallow to moderate depths, there is a possibility that warm to hot waters are mixing with cooler surface water at depth. As a routine operation temperatures should be measured in exploratory bore holes to determine whether unusually warm or hot conditions exist at depth.

Water circulating downward from the earth's surface commonly has the same temperature as the immediately adjacent rocks, and the temperature at depth can be estimated from the *geothermal gradient.* The geothermal gradient expresses the increase in temperature with depth. The *reciprocal gradient* is the number of feet or meters of vertical penetration of the subsurface that will result in an increase in temperature of one degree Fahrenheit or Centigrade. In geologically stable regions of the earth the reciprocal gradient is about 60–80 m per increase in rock temperature of 1°C (approx. 100–145 ft/°F). In volcanic regions characterized by igneous activity at the present time or in the recent geologic past the reciprocal gradient commonly is near 10–15 m/°C (18–27 ft./°F). A world-wide average is 30–35 m/°C. Except in volcanic regions or regions of unusual crustal activity it may be expected that water (and rock temperatures) in most tunnels will be within a reasonable range.

The geothermal gradient differs with the attitude of layering in rocks in a region with a laterally uniform thermal history because of a greater thermal conductivity parallel to layering than across it. In general, temperatures increase more rapidly with depth in flat lying than in steeply dipping layered rocks by a factor of 1.2–1.4.

In estimating increase of temperature with depth based on an assumed or known gradient temperatures at and near the surface to a depth of the order of 20–25 m (60–80 ft.) may be safely ignored under most circumstances. Rock temperatures at the base of the surface zone commonly are 4–10°C (40–50°F), temperatures that may be used as starting points for calculation of temperatures at depth with a known or assumed gradient.

Calculations based on an average geothermal gradient are of little use when thermal waters in underground openings are localized by fracture zones, a common circumstance. When thermal waters are confined to a fracture zone it is quite likely that the rock temperatures adjacent to the fracture zone are the result of flow of heat from hot water into the rock, and the temperature of the water does not depend on rock temperature.

In many tunnels flows of large volumes of hot water were totally unexpected and were encountered in "hot spots" with temperatures considerably above the prevailing rock temperatures. The task of the geologist is formidable in predicting such conditions, but using every instinct that he possesses, he should attempt to do so.

FRACTURE PERMEABILITY

Fracture permeability is secondary permeability resulting from development of faults and joints in rocks of variable original permeability. Very-heavy water flows sometimes are encountered underground in fractures, especially in open interconnected fractures. In addition, because open fractures may offer relatively slight frictional resistance to flow, pressures at the point of intersection with open fractures may be very high and in some instances may approach the hydrostatic pressure associated with a standing column of water (0.433 lbs inch^{-2} ft.$^{-1}$ at 4°C). Fractures guide the movement of underground waters, and in some rocks, especially in limestone, fractures are enlarged by dissolution of rock adjacent to the fractures, a subject that will be discussed in a later section of this chapter.

The various kinds of fractures that are capable of controlling the flow of underground water have been summarized in Chapter 5, and no effort is made in the present discussion to characterize fractures systematically. Instead, attention is given to some of the conditions associated with faults that are likely to produce

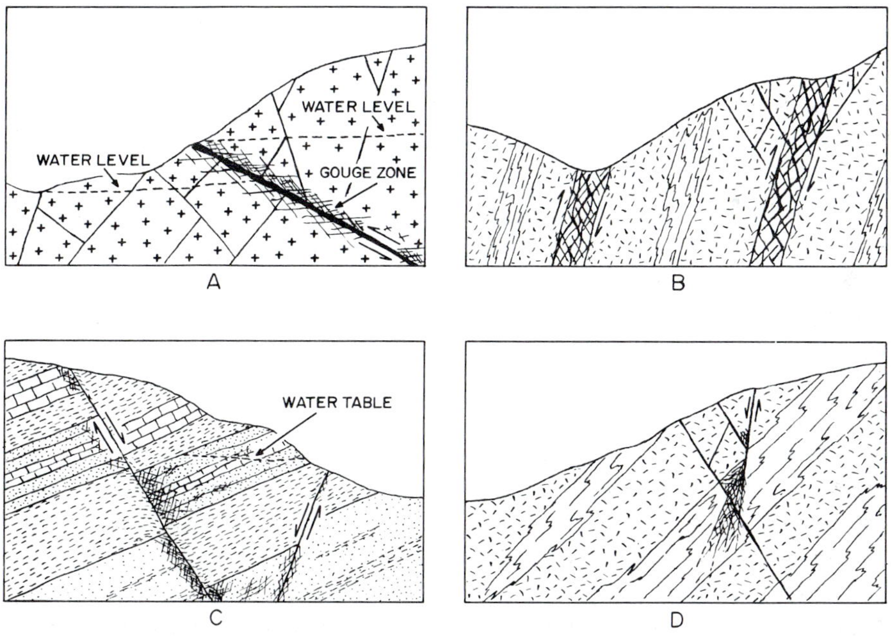

Fig. 56. Geologic conditions leading to underground accumulation of bodies of trapped water. A. Ground water is stored in fractured rock above a gouge-filled thrust fault in crystalline rocks. B. Wide, intricately fractured shear zones contain abundant water. C. Water is trapped in highly fractured, brittle sandstone and limestone adjacent to a large fault. Shale provides impermeable barriers. D. A sharp fault in granite intersects schist and gneiss and horse-tails into an intricate pattern of intersecting fractures.

heavy, continuing flows of underground water when encountered in underground openings.

Two kinds of underground water flows can be distinguished: (1) flows in fractures connecting with the earth's surface and fed by water from streams, standing bodies of water, or water-soaked surficial deposits, and (2) flows of stored or trapped water which is replenished relatively slowly by water from the earth's surface.

Flows of surface water into underground openings can be anticipated in some situations, especially when a tunnel passes beneath a stream valley localized along a shear zone, or beneath a lake in a region of extensive bedrock fracturing. Prediction of the magnitude and location of bodies of stored or trapped water is not so easy, and must be based on a detailed knowledge of local geology, especially the distribution of faults capable of generating extensive fracture permeability.

Fig. 56 indicates a few geologic conditions that might lead to accumulation of extensive bodies of stored or trapped water.

PERMEABILITY IN LIMESTONES

Limestone consists predominantly of calcium carbonate and is particularly susceptible to dissolution by circulating ground water, and to chemical reactions leading to changes in bulk composition. Original porosity and permeability in limestones depends on origin and ranges from a minimum in very fine-grained compact accumulations (micrite) to a maximum in poorly consolidated accumulations of oolites or fossil remains of organisms. Most limestones buried at moderate to considerable depths by superjacent sedimentary layers are converted to coherent, brittle rocks with a notable reduction in porosity by recrystallization and by deposition of cement in pore spaces.

Many limestones contain some dolomite and all compositional gradations exist between pure limestone and rocks that consist entirely of dolomite, so-called *dolostones.* Dolomite can be precipitated as an original mineral, but in many natural occurrences, the dolomite appears to have developed in limestone slightly or considerably later than the time of initial deposition of the limestone. Accordingly, much dolomite is found in *dolomitized limestones.*

Dolomitization of limestones after they have been deposited and consolidated by burial is a consequence of reaction with ground waters of ultimate surface origin or by reaction with hydrothermal solutions of deep-seated origin. In many instances, particularly where hydrothermal solutions have been active, dolomitization is accompanied by a marked increase in porosity and permeability of the host rock, and interconnecting solution cavities provide channelways for relatively easy circulation of ground water. Not uncommonly hydrothermal dolomitization is accompanied by extensive silicification which produces very brittle rocks that fracture easily under local or regional stresses and develop high-permeability aquifers.

Brittleness in compact limestones promotes development of fracture permeability, which, with permeability parallel to bedding planes, enables easy circulation of ground waters. Although limestone dissolves to a slight extent in pure water, the presence of carbon dioxide in ground water spectacularly increases its capacity to dissolve limestone. Carbon dioxide is supplied to ground water mainly by the atmosphere, although many thermal waters contain carbon dioxide of deep-seated origin. Large-scale dissolution of limestone requires a steady replenishment of the carbon dioxide content of ground water, an observation supporting the fact that limestone dissolution is at a maximum where geologic conditions favor continued underground circulation of waters of surface origin. Provided that a supply of carbon dioxide is available, dissolution occurs above the ground-water table as well as below it in actively circulating systems.

Above the ground-water table a complex network of caverns and narrow irregu-

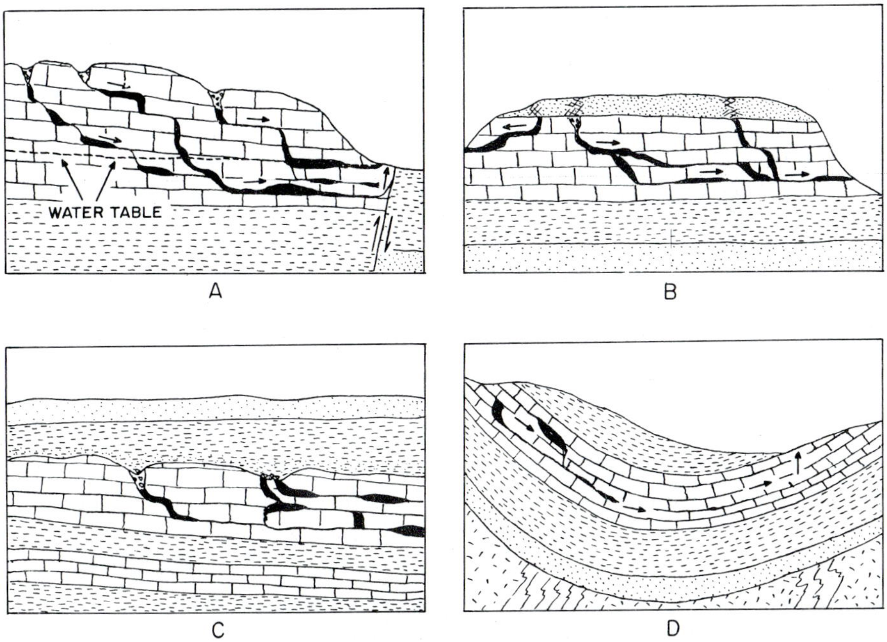

Fig. 57. Origin and distribution of dissolution openings in limestone in selected examples. A. (Caverns (black) develop above and below ground-water table. Sinkholes (karsts) formed by collapse of caverns at surface. B. Caverns develop below a permeable sandstone capping. C. Ancient karst topography is buried by later sedimentary layers. There is no surface indication of the presence of caverns in the limestone. D. Cavernous limestone develops down dip in a folded limestone by circulating ground water, containing atmospheric carbon dioxide.

lar tunnels control the flow of underground rivers and impound lakes. Below the ground-water table caverns and fissures are filled with stagnant or actively moving water which yields torrents when encountered during excavation of underground openings.

Channelways of dissolution in limestone generally are very irregularly and intricately interconnected, and prediction of their locations is virtually impossible. Clues to the possible location of caverns sometimes are provided by a surface study of fracture systems which may localize underground circulation. The probable existence of solution caverns, wherever they may be located at depth, is indicated by sinkholes (karsts) at the surface over a limestone and by heavy flows of water issuing from springs at or near surface exposures of a limestone.

A few examples indicating the origin and distribution of dissolution openings in limestone are shown in Fig. 57.

SQUEEZING, SWELLING, AND RUNNING GROUND

Among the most feared hazards in tunneling operations are altered and/or fragmented rock masses that swell, squeeze, or flow into tunnel openings. Pressures generated by such materials are capable of causing failure of heavy supports, and containment may be a major problem. Almost invariably squeezing and swelling are associated with original pore water or with water introduced into the materials from adjacent water-bearing fractures or from humid tunnel atmospheres, or both.

Technically, *squeezing ground* is weak, plastic material that moves into a tunnel opening because of the weight of superjacent rocks, and *swelling ground* is material that is displaced into a tunnel opening because of an increase in volume by absorption and adsorption of water. Under many circumstances a clear distinction cannot be made between swelling and squeezing ground because introduction of even small amounts of water increases the tendency toward plastic behavior in squeezing ground.

A valid distinction between squeezing and swelling ground is based on the time of first noticeable movement of plastically-behaving materials into the tunnel opening. In general, squeezing ground starts visible movement almost as soon as it is exposed, and it is apparent that the movement is caused by unbalanced stresses. Swelling ground, on the other hand, requires addition of water before it expands, a process that is relatively slow. Accordingly, swelling may not be noticed until considerable time has passed.

Of course, swelling pressures commonly are converted into squeezing pressures. Moreover, as swelling ground expands, it also tends to become more plastic, so that it will deform under unbalanced loads in the same manner as ground that initially behaves as squeezing ground. There is no inconsistency in classifying some materials in underground openings as swelling, squeezing ground.

Running ground is water-soaked fragmented rock often containing larger rock fragments embedded in a matrix of fine grained material that has no strength and flows into a tunnel as a slurry. Running ground in some instances results from diversion of ground water into initially relatively dry crushed aggregates along channelways accidentally created by the tunneling operation or by rearrangement of underground water-flow patterns by grouting or tunnel lining.

Causes and mechanics of swelling in rocks

Swelling in mineral aggregates is caused by one or a combination of processes including (1) adsorption of films of water attracted and held by surface forces of very small mineral particles, (2) absorption of loosely held water by layer silicates such as montmorillonite, (3) hydration, and (4) slight expansion of pore water as a consequence of release of confining pressure.

With the exception of anhydrite ($CaSO_4$), which swells because of chemical incorporation of water to form gypsum ($CaSO_4 \cdot 2H_2O$), swelling is most pronounced in rocks which contain abundant clay minerals or clay-size particles of other

Fig. 58. Swelling ground in sheared, thoroughly argillized schist and gneiss. Vasquez Tunnel, Colorado. (Photo by Denver Board of Water Commissioners.)

minerals (Fig. 58). Expanding clay minerals, especially montmorillonite and mixed-layer clay minerals containing montmorillonite, absorb water and, when abundant, cause spectacular swelling of the materials in which they are present. Swelling of very fine-grained rocks that do not contain montmorillonite is less pronounced and results mainly from adsorption of water on the surfaces of mineral particles.

Swelling is commonly a slow process, primarily because of the fine grain of the materials that are prone to swelling. The permeability of such materials is low, and penetration by water of any origin takes considerable time. Swelling promoted by ground water circulating along fractures in adjacent more competent materials is easily identified, but swelling as a consequence of permeation by moisture from the tunnel atmosphere or from other causes is a more subtle process.

Theoretically swelling can be prevented by sealing off materials prone to swelling by grouting or diverting channelways of flow of underground water and by covering exposures of swelling ground with impermeable concrete or asphalt to prevent access of atmospheric water. However, in practice, these measures rarely prevent swelling because of the presence of pore water under pressure in the swelling material.

The natural processes which generate very fine-grained clay-rich materials generally require abundant water, as for example, in weathering, deposition of argillaceous sediments, and in hydrothermal alteration. A consequence is a residue of pore water which is confined under pressure approximately equal to the pressure of the overlying column of rocks. Release of confining pressure by a tunnel excavaton initiates a slow movement of pore water toward the tunnel opening and provides a steady supply of water contributing to the swelling of materials adjacent to the tunnel opening.

The movement of pore water is of the same kind as that observed in saturated clay dam embankments during draw-down of the reservoir behind the embankment.

Two factors operate to cause swelling gradually to diminish:

(1) The supply of pore water migrating toward the tunnel opening is reduced with time and eventually a stable pressure gradient is established.

(2) Swelling resulting from absorption and adsorption of water from the tunnel atmosphere produces a layer of expanded materials which shields unexpanded materials from tunnel moisture.

Equilibrium may not be reached for several weeks or months, and finding a solution to the problem of containment of swelling rocks until equilibrium has been attained may be very difficult. In any event, swelling is a phenomenon localized by the tunnel opening and affects materials outside the opening for only a limited distance, commonly measured in feet or tens of feet.

A variety of natural materials exhibits a swelling tendency. Some common materials that swell are described in the following paragraphs.

Shales, claystones, and *mudstones.* These rocks are very fine-grained and exhibit a wide range of mineralogic compositions. Notable common mineral components other than the clay minerals are quartz, calcite, dolomite, carbonaceous substances, and iron oxides. The swelling tendency is directly proportional to the amount of clay minerals, especially montmorillonite or intergrowths of montmorillonite with other clay minerals. Shales, claystones, and mudstones rich in clay minerals and deficient in other mineral components generally are soft and possess slight strength. Swelling may be very slow because of impermeability, especially in rocks encountered at shallow depths where pore pressures are not high.

Altered pyroclastic deposits. Buried deposits of volcanic ash commonly have been converted to *bentonite,* a material rich in swelling clays, especially montmorillonite. Expansion by absorption of water together with adsorption on small particles of other minerals commonly develops spectacular swelling.

Fault gouge. Fault gouge is capable of incorporating considerable adsorbed water. If clay minerals are present they contribute to the swelling tendency.

Weathered rocks. Weathering commonly produces clay minerals at the expense of original minerals. In portions of weathered zones where leaching of alkalies and alkaline earths is incomplete, water-absorbing montmorillonite or mixed-layer minerals containing montmorillonite are likely to be present. Abundant kaolinite, illite, sericite, or fine-grained chlorite also promote swelling by adsorption of films of water.

Hydrothermally altered rocks. Hydrothermal alteration commonly is localized by channelways of circulation, and the alteration spreads out from faults, joints, and other kinds of interconnecting flow channels, such as brecciated tops of lava flows or bedding planes. Alteration that produces swelling ground is characterized by introduction or formation in place of very fine-grained aggregates of clay minerals, including fine-grained chlorite and sericite. Silicification and carbonatization, common types or hydrothermal alteration, tend to improve the quality of rock, and an abundance of introduced quartz or carbonates tends to reduce or eliminate the swelling tendency.

Anhydrite-bearing rocks. Anhydrite ($CaSO_4$) commonly is associated with reddish shales, siltstones, and mudstones and, in many occurrences, with rock salt and limestone. Anhydrite exposed to water is very slowly converted to gypsum ($CaSO_4 \cdot 2H_2O$) and generates considerable swelling pressures. However, the swelling may be prevented by denying access of water to exposures.

Swelling pressures of clay-bearing rocks

Many devices are available for measuring volume increases of swelling ground when they are brought into contact with water and for measuring pressures generated by swelling. A first clue to the existence of swelling tendency may be obtained by placing samples of suspected material in water and observing the effects over a

TABLE XV

Swelling properties of gouge and altered rock, Roberts Tunnel, Colorado

Description	*Final heave* (% vol. change)	*Final swell pressure* (p.s.f.)
I. Gouge consisting predominantly of crushed wall rock with minor to moderate clay-mineral alteration	0.5 –10.0	300– 4,250
II. Gouge containing abundant clay minerals, with predominant montmorillonite, and smaller amounts of kaolinite and sericite. Clay minerals are of hydrothermal origin.	10.25–18.5	4,500–11,800
III. Hydrothermal altered wall rock with minor to moderate amounts of clay minerals. Some samples show extensive hydrothermal silification.	9.0 –10.0	3,900– 4,250
IV. Hydrothermally altered wall rock with abundant clay minerals, predominantly montmorillonite, but with some kaolinite, sericite, and chlorite	13.75–25.0	8,000–12,000

period of time. Swelling materials gradually expand and disintegrate in water over periods of a few minutes to several days or weeks. Increases in volume of more than 25% are not unusual.

The author has obtained excellent and consistent results in the study of volume changes and swelling pressures with an instrument designed by Lambe (1960) for study of expansive soils. With suitable modifications of procedure the instrument may be adapted for use with crushed, altered rocks. Some notion of the magnitudes of swelling pressures in fault gouge and in associated altered wall rocks is provided in a summary in Table XV of a study, using Lambe's techniques, of the Harold D. Roberts Tunnel, Colorado (Wahlstrom et al., 1968). All samples were from portions of the tunnel in crystalline rocks, either ancient (Precambrian) metamorphic gneisses and schists or from the interior of a stock of quartz monzonite. The values reported in Table XV are in two categories, *final heave*, and *final swell pressure*. Final heave measures the volume increase of an appropriately compacted sample placed in a consolidometer under an initial pressure of 200 lbs/sq. ft. and placed in contact with water. The reported results express change in height divided by initial height when equilibrium has been reached. Final swell pressure is measured in the consolidometer in a sample subjected to an initial pressure of 200 lbs/sq. ft. Pressures are measured as water is added while the initial height of the sample is maintained.

In the Roberts Tunnel swelling ground with high swelling pressures, especially in the quartz monzonite, caused dislocation and bending of heavy steel supports and required extensive installation of additional supports several weeks to several months after the rock was first encounterd in the tunnel heading. High swelling pressures confined to gouge in narrow faults proved to be of minor concern.

Squeezing ground

Squeezing ground is plastic material that moves into an underground opening primarily because of pressures exerted by loads of overlying rock (Fig. 59 and 60). Although a clear distinction is not always possible, squeezing ground differs from swelling ground in that it undergoes no appreciable volume increase owing to penetration by water. Of course, ingress of water, even in small amounts, promotes plastic behavior and contributes to the easy movement of squeezing material.

Pressures causing squeezing at times approximate dead-load rock pressures of overlying rock columns, and are sufficient to cause failure of heavy steel supports

Fig. 59. Squeezing ground in incompetent sheared granite and gneiss. Roberts Tunnel, Colorado. Note steel rail spiling and deformed steel ribs. (Photo by Denver Board of Water Commissioners.)

Fig. 60. Squeezing ground in altered schist and gneiss. Vasquez Tunnel, Colorado. Note crushed and splintered lagging. (Photo by Denver Board of Water Commissioners.)

or concrete tunnel linings. Unlike swelling ground, movements in squeezing ground involve materials outside of the tunnel opening, and the volumes of material that have the potential of moving into the tunnel opening may be very large.

Squeezing occurs when the load pressure on a plastic or semiplastic material exceeds a critical value beyond which there is permanent strain. That is, most squeezing materials have some initial inherent strength and, up to the limit of this strength, resist dislocation. Accordingly, whether a particular material squeezes or not depends on the local conditions. Moreover, the rate of squeezing is a function of the characteristics of the material and the magnitude and directions of the forces that are being exerted to cause the squeezing.

Several kinds of natural materials are prone to squeezing. Many of these materials are rich in clay minerals and in the presence of water also tend to swell. At shallow depths, where load pressures are low, materials most susceptible to squeezing contain abundant pore water, and have very-low initial strength. Examples are fault gouge, highly altered pyroclastic deposits containing expanded clay minerals, closely sheared and altered micaceous rocks, and poorly consolidated, soft mudstones and claystones. At greater depths fault gouge and closely sheared, altered rocks, especially rocks containing abundant micaceous minerals and clay minerals, are among materials most likely to squeeze. Undisturbed sedimentary shales, mudstones, and claystones at moderate to considerable depth generally have been sufficiently compacted by the load of overlying rocks to have developed enough strength to resist squeezing dislocation. In addition, confinement of semiplastic materials under considerable rock loads tends to increase beyond the periphery of the tunnel opening, as in triaxial compression tests, so that the threshold level of permanent strain is higher than at shallow depths where confining pressures are lower.

As would be expected, the rate of squeezing in materials of similar mineralogy and grain size generally is directly proportional to the water content. All gradations exist between materials that move at a visible rate and stiff materials that squeeze

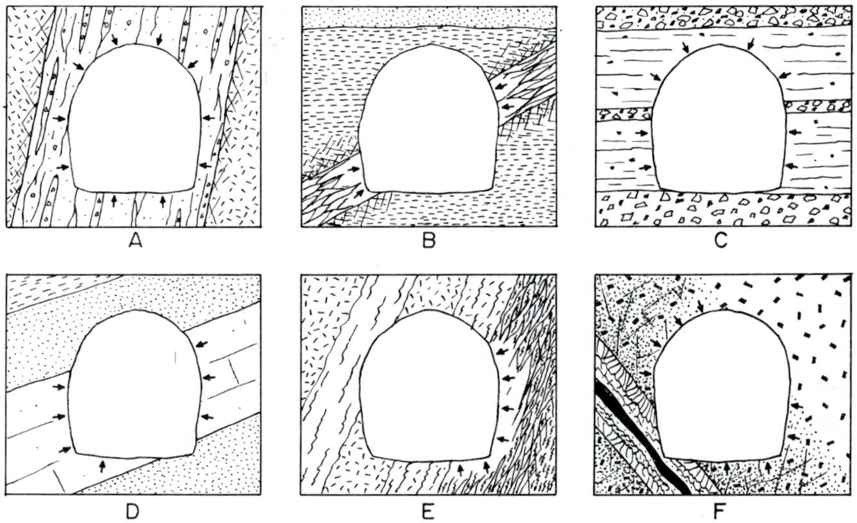

Fig. 61. Some conditions promoting squeezing ground in tunnels. Arrows indicate direction of movement of squeezing ground into tunnel opening.
A. Tunnel is in a wide shear zone in granite. Rock has been thoroughly crushed and consists of a gouge matrix containing lenses of angular rock fragments. Gouge is essentially impermeable to penetration by water but is damp.
B. Flat fault in brittle shale has produced gouge containing braided pattern of shears. Swelling is localized at exposures of fault.
C. Layer of altered volcanic ash (bentonite) contains abundant expanded montmorillonite. Load pressure of overlying rocks causes squeezing.
D. A soft, poorly compacted mudstone layer is squeezed into tunnel by weight of overlying rocks.
E. Closely sheared, altered biotite schist contains abundant chlorite and clay minerals and squeezes into tunnel opening.
F.An extensively altered fine-grained porphyry adjacent to a vein tends to squeeze into tunnel opening.

very slowly. In any event, movement tends to cause work hardening that may be the result of slow desiccation as well as reorientation of mineral particles into more stable configurations. It is unusual for squeezing materials to continue movement after a period of time of the order of several weeks to several months, and the movement of squeezing materials rarely extends outside of the tunnel opening for a distance in excess of several tens of feet, except in very large tunnels.

Some examples of geologic conditions that promote squeezing are indicated in Fig. 61.

Running ground

Examples of dry running ground are known, but are rare. In some mining districts dolomitization of limestone has produced loose sand-like accumulations of crystals that flow freely into openings when encountered. A notable example is in the Gilman mining district in Colorado. Also, at times very poorly cemented sandstones or unconsolidated fault breccia in brittle rocks, such as quartzite, are capable of moving freely into underground openings.

Fig. 62. Breastboards installed to contain highly incompetent running ground in closely fractured, water-bearing granite and gneiss. Roberts Tunnel, Colorado. (Photo by Denver Board of Water Commissioners.)

Fig. 63. Running ground in closely sheared shale. Roberts Tunnel, Colorado. Brest-boards are installed as containment. (Photo by Denver Board of Water Commissioners.)

Typically, running ground is water-saturated, and consists of intensively crushed rock with a high degree of permeability (Fig. 62 and 63). Gouge may be present, but not in sufficient amounts to reduce permeability to a notable extent. Clay minerals, if present, promote weakness and contribute to the easy movement of the material. Very commonly running ground flows into a tunnel opening looking very much like mud slurry mixed with small and large, angular rock fragments.

Running ground generally is associated with faults and in particular with zones of highly-fractured rock in the footwall or hanging wall of a fault. In some instances, definite surfaces of fault dislocation are identified with difficulty and, instead, the running material before dislocation appears to consist of a loose, jumbled aggregate with indefinite gradational boundaries.

Initial pressures generated by running ground may be sufficient to cause failure of very heavy supports.

TUNNELING IN SQUEEZING, SWELLING, AND RUNNING GROUND

The literature on tunneling records many spectacular and very costly experiences with squeezing, swelling, or running ground. Except for the problems created by inflow of large volumes of water, ground that squeezes, swells, or runs should not be considered as a major persistent geologic hazard in tunneling. The rate at which squeezing and swelling involve rock masses differs widely, depending on local geologic conditions, but, running ground, when encountered, always presents a problem that must be coped with immediately. Whatever the character of the material it is generally safe to predict that, given sufficient time, natural processes will tend to bring the material into a condition of approximate stabilization, and movement will slow down and eventually stop. In many situations heavy pressures on supports initially installed to contain the material gradually diminish in intensity over short to long periods of time.

There is considerable difference of opinion as to the appropriate method for dealing with swelling and squeezing ground in underground openings. Simply stated, there are two alternatives: (1) install supports sufficiently strong to contain the swelling or squeezing ground, a somewhat unsatisfactory and frustrating procedure when extensive intervals of ground with high swelling or squeezing pressures are encountered, or (2) allow the swelling or squeezing material to move under controlled conditions into the tunnel opening with the logical expectation that the movement will diminish and eventually stop. The writer prefers the second alternative, because a stabilized swelling or squeezing condition, once attained, is no longer a matter of concern, whereas a contained, actively swelling or squeezing material poses a continued threat to the integrity of the confining supports that may lead to failure.

In many situations failure of concrete tunnel linings can be avoided either by delaying placement of the lining until squeezing or swelling have become negligible or by leaving openings in the lining that will allow pressure relief until stabilization is finally attained.

When heavy, wet, running ground is encountered, there is a strong temptation to stop the tunneling operation, perhaps with the installation of perforated pipe, and allow the material to drain. Generally this procedure is not feasible because of time limitations imposed by construction schedules. Nevertheless, many battles are won in tunnel sections of running ground, not so much because of the installation of heavy supports and utilization of techniques of consolidation such as grouting, but because natural drainage, accompanied by gravity removal of fine-grained materials, gradually decreases the running tendency in the residual fragmental rock material. Also adjustments in adjacent rocks tend to fill openings left by running ground and may improve stability. In any event, allowing controlled flow of running ground into a tunnel opening at times may have many short and long term advantages over attempts to contain the ground completely by timber, steel supports, or concrete lining before a considerable degree of stabilization has been reached. As a general rule, installation of totally confining supports should be delayed as long as possible under all conditions where there are short-term or long-term dislocations of materials into a tunnel opening, whatever the intended use of the tunnel may be.

Special consideration needs to be given to geologic conditions in water tunnels, either gravity-flow or pressure tunnels. Clearly, every effort should be made to seal off masses of rock that squeeze, swell, or run before the tunnel is put to its intended use. During the tunneling operation careful examinations should be made of exposures of dry or moist materials, especially in fault zones, that might be converted to running or squeezing ground by slow penetration of water while the tunnel is in operation.

A particularly dangerous situation is created in pressure tunnels where water under a head penetrates initially dry, altered or fractured rock. Variations in pressure, especially decreases in pressure, create a highly unstable condition because of a build-up of pore pressures resulting from slow penetration of water under the pressure prevailing in the tunnel. To prevent ultimate collapse of such material, every effort should be made to seal it off by impermeable, crack-free high-strength tunnel lining.

Several examples of delayed collapse of fractured rock in pressure tunnels have been cited by Detzlhofer (1968).

TESTING OF PERMEABILITY OF ROCKS

The testing of permeability in rocks that will be encountered in tunneling operations is much more complex than testing of layers or blankets of unconsolidated deposits of surficial origin. Although some rocks, such as poorly cemented, clean sandstones, may have higher than average original permeabilities, such rocks rarely present problems in tunnels. Instead the concern is chiefly with secondary fracture permeability or permeability resulting from dissolution, as in limestones.

Some information regarding secondary permeability is available in areas where springs issue at the surface, but in most circumstances estimation of rock permeability must be based on surface geological and geophysical investigations and on information derived from exploratory drill holes. Drill holes provide linear samples, and, in most instances, there is no assurance that a particular hole will yield even an approximately accurate picture of fracture and/or dissolution permeability in a rock mass. Nevertheless, careful logs should be kept of drill holes and special note should be made of water losses or grains during drilling. In addition, tests should be made at frequent intervals measuring losses of water pumped into the hole under pressure. Extensive losses of water suggest the distinct possibility that heavy flows of water will be encountered in the vicinity of the drill hole as the tunnel is driven. However, the reverse is not necessarily true, that is, low losses prove only that the drill hole did not encounter any notable secondary permeability, although such permeability may be present adjacent to the hole.

Quantitative estimations of secondary permeability and the amount of water that will be encountered in a tunnel generally should not be attempted unless geologic conditions are relatively simple and well known. Instead, predictions should be based on the best possible subjective estimate of the maximum flows of water that might be encountered, so that provisions can be made for grouting and installation of adequate pumping facilities. The best safeguard against heavy water flows is drilling feeler holes well in advance of the tunnel heading, so that heavy flows of water can be located before they are intersected by the tunnel.

GASES IN ROCKS

A variety of gases of natural origin have been encountered in tunnels in amounts and proportions that are not characteristic of the earth's atmosphere. Among these are carbon dioxide, "methane", sulfur dioxide, hydrogen sulfide, and, rarely, hydrogen. "Dead air" has essentially the composition of air from which oxygen has been removed by oxidation reactions with iron compounds and sulfides. Many devices are available for monitoring tunnel air for undesirable constituents.

Carbon dioxide is a gas with a density approximately 1.5 times that of air and tends to collect in lowest available spaces. Abundant carbon dioxide, in amounts in excess of 15–25% of the tunnel atmosphere, is lethal. Combined with water it produces carbonic acid which is mildly corrosive in contact with metals and may cause deterioration of concrete lining.

Carbon dioxide results from the oxidation of carbonaceous materials, but, in addition, it is a common and abundant constituent of warm to hot waters of deep-seated origin. Especially large amounts of carbon dioxide can be expected in regions of relatively recent igneous activity.

"*Methane*", also called *marsh gas,* is a simple hydrocarbon with the formula CH_4. Much so-called methane contains other hydrocarbons. The gas is nontoxic, has a density only about half that of air, and in concentrations in excess of about 5% forms a highly-explosive mixture. An explosion and fire caused by methane combustion gas from shale created a major disaster during the driving of the Great Appenine tunnel between Italy and Switzerland. The gas has also contributed to major explosions in many coal-mining operations.

Methane is most abundant in the vicinity of coal and petroleum deposits, but exists also in considerable quantities in other sedimentary environments, where it has formed by the slow decay of bituminous materials of organic origin. The gas, when confined by impermeable barriers, can move for great distances laterally and vertically in fissured and porous rocks to develop accumulations in sometimes unsuspected locations.

Sulfur dioxide and *hydrogen sulfide* are easily detected by their distinctive odors. Sulfur dioxide most frequently is a consequence of oxidation of sulfur or sulfides in sedimentary accumulations or in hydrothermal mineral deposits containing concentrated or disseminated sulfides. Hydrogen sulfide is a by-product of the decay of sulfur-bearing organic materials, sulfides of inorganic origin, and is a constituent of many thermal waters of deep-seated, presumably igneous, origin. Hydrogen sulfide in concentrations in air of less than 1% is lethal, but, because its odor gives an early warning of its presence, removal by adequate ventilation can be initiated to prevent build-up of lethal concentrations.

REFERENCES

Detzlhofer, H., 1968. Verbrüche in Druckstollen, In: L. Muller (Editor), *Felsmechanik und Ingenieurgeologie.* Suppl. IV. Springer, Berlin–Heidelberg–Göttingen, pp. 158–180.

Lambe, T.W., 1960. The character and indentification of expansive soils. *Fed. Housing Adm., Tech., Stud. Rep.,*FHA-701, 46 pp.

Wahlstrom, E.E., Robinson, C.S. and Nichols, T.C., 1968. Swelling of rocks in faults in the Roberts Tunnel, Colorado. *Eng. Geol. Case Hist. – Geol. Soc. Am.,* 6: 83–89.

Chapter 7

GEOLOGICAL INVESTIGATION OF PROPOSED TUNNEL LOCATIONS

INTRODUCTION

Geological investigation prior to the design and construction of a tunnel (or any excavated underground facility) requires careful attention to all geological features, gross and detailed, that might have any bearing on the rate of progress and the costs of construction, whatever method of tunneling is employed.

This chapter summarizes and makes appraisals of various standard techniques and procedures used in geological investigations of proposed tunnel locations. The reader will soon realize, as he reviews the various methods and techniques, that geology is not and can not be an exact science, although it employs many instruments that yield precise readings. Projection to tunnel elevation of geological conditions observed on the earth's surface or logged in core holes, together with interpretation of geophysical measurements, especially in areas of extensive surface cover and/or complex bedrock geology, requires exercise of careful judgment based on extensive experience as to what is certain, what is probable, what is possible, and, finally, estimation of what is highly unlikely.

The treatment in this chapter presupposes familiarity with the text materials in preceding chapters.

GEOMORPHOLOGY AND SUBSURFACE GEOLOGY

Geomorphology is concerned with the natural processes that shape the surface of the earth. The processes include large and small dislocations of the rocks of the earth's crust, weathering and erosion by water, erosion by wind and ice, and transportation and deposition of the waste products of weathering and erosion. Much of the earth's surface is covered by water. The remainder, constituting the land areas, demonstrates relief ranging from almost negligible in extensive flat areas to extreme in rugged mountainous areas.

Land forms observed at the earth's surface generally are the culmination of complex interaction of processes which, in a broad characterization, can be classified as *endogenous* and *exogenous*. Endogenous processes include those that result in local to regional uplift or sinking of segments of the earth's crust accompanied by folding and faulting, elevation or depression of very large land areas (epeirogenic movements), and igneous activity, both intrusive and extrusive. Exogenous processes are active at the earth's surface and tend to destroy or bury land forms created by endogenous processes.

The manner in which exogenous processes operate to modify or create land forms at the present time depends to a large extent on the climate. Weathering and erosion in warm-humid climates provides a marked contrast to weathering and erosion in cold-humid, glacial, or arid climates. However, many landforms observed today obviously developed some or most of their characteristics at some time in the geologic past, when the climate was considerably different from the present climate.

Much of the surface of the land is covered by vegetation or the products of weathering and erosion, which have been removed small to great distances from the point of origin. The task of the geologist who is attempting to determine the bedrock geology along a proposed tunnel location frequently is made very difficult by the presence of extensive surficial deposits or vegetation which blanket rock exposures, and he must rely to a large extent on an intelligent interpretation of the origin and significance of various topographic features to arrive at an assessment of rock conditions at depth.

Broad flat areas of low relief, generally completely covered by surficial deposits, do not permit interpretation of details of geology of the bedrock by examination of surface features. However, in many regions, particularly in those regions of interest to a tunnel geologist, the topographic relief is sufficient and the shapes of the landforms are characteristic enough that much can be learned about the geology of the bedrock, even though very few rock exposures may be present.

It is with these conditions in mind that the classification of landforms as they are related to erosional or depositional history and subsurface geology was prepared and summarized in Table XVI. This classification is intended for use by the field geologist or a geologist working with aerial photographs or images obtained by remote sensing in his efforts to correlate shapes of landforms with various subsurface geologic features. Illustrations of several landforms and related subsurface geology are shown in Fig. 64–70.

Much can be surmised about the structure of bedrock by inspection of drainage patterns on aerial photographs and topographic maps. A few examples of distinctive drainage patterns are shown in Fig. 71.

TABLE XVI

Outline classification of landforms

I. Landforms developed by erosion of layered rocks.

A. Landforms developed in tilted or folded layered rocks, whether igneous (extrusive), sedimentary, or metamorphic, and formed by differential erosion in rocks with alternating hard and soft layers.

1. Layered rocks form domes as a consequence of forceful intrusion of magma or evaporitic material (rock salt, anhydrite) from below. Process sometimes called *diapirism.*

   Landscape characterized by circular or concentric arcuate hogbacks or low ridges.

   Examples: Eroded domical structures formed above salt intrusions, igneous stocks, or laccoliths (Fig. 64).

2. Layered rocks assume domical configuration by differential compaction of sediments over topographic prominences in basement rocks.

   Erosion produces circular or concentric arcuate hogbacks or low ridges.

3. Layered rocks form synclines, anticlines, or monoclines. Resistant rocks hold up sharp hogbacks or low ridges, either linear or curvilinear.

   Examples: Eroded anticlines and synclines with horizontal fold axes (Fig. 65), anticlines and synclines with plunging (nonhorizontal) hinge lines (Fig. 66), and tilted rocks with a single direction of dip (Fig. 67).

B. Landforms associated with essentially horizontal layered rocks or weathered zones. Topographic prominences commonly held up by caprock. Landscape characterized by steep-walled cliffs, bluffs, and scarps developed in resistant rocks.

1. Sedimentary, igneous (extrusive), or metamorphic rocks produce plateaus, mesas, or buttes.

2. Weathering has produced mantles on basement rocks which resist erosion.

   Examples: Silica-rich mantles of *silcrete* or *duricrust* mantles which cap mesas, bluffs, or plateaus. Sometimes iron-rich mantles produce the same features.

II. Landforms are associated with intrusive igneous bodies which have been exposed at the earth's surface by erosion or faulting accompanied by erosion. Topographic expression depends on greater resistance of igneous rocks to erosion as compared with country rock.

A. Topographic prominences associated with discordant igneous intrusions.

1. Landforms developed over cross-cutting dikes which produce essentially linear topographic features.

2. Domical topographic features produced by denudation of igneous stocks.

B, Landforms developed on concordant igneous intrusions.

1. Topographic features including hogbacks, or low ridges in tilted rocks, and caprocks and cliffs formed in horizontal-layered rocks by sills.

2. Domical topographic features produced by erosion and exposure of laccoliths.

III. Landforms formed by extrusive igneous activity which has produced generally layered rocks consisting of lava flows and/or pyroclastic materials (Fig. 13).

A. Eruption of volcanic material, usually of basaltic composition, is localized along a fault or a succession of faults to produce fissure flows which may spread laterally for great distances.

   Examples: plateau and shield basalt flows.

TABLE XVI (continued)

B. Eruption is localized by a volcanic neck and produces volcanic cones, lava domes, and lava spines.

IV. Landforms controlled by fault dislocations and associated joint sets.
- A. Relative elevation and depression of adjacent blocks dislocated by faulting is preserved more or less and is expressed topographically.
  - 1. Landform is associated with a single fault or fault trend.
    - a. Fault scarp or fault-line scarp which is relatively straight for steep faults (Fig. 68A,B), generally curved for flat faults (Fig. 68C), or circular as in volcanic sinks and calderas (Fig. 68F).
  - 2. Landform is associated with paired, essentially parallel faults.
    - a. Horst (Fig. 68D).
    - b. Graben (Fig. 68E).
- B. Relative elevation and depression of adjacent fault blocks cannot be determined from configuration of landform.
  - 1. Landforms in extensively planated areas. Valleys along fault zones or joint sets are cut below surfaces of low relief.
  - 2. Landforms are valleys following faults in regions of crystalline rocks where relative displacement along faults cannot be determined because of lack of structural controls provided by layered rocks (Fig. 69).
- C. Landforms developed along mineralized, especially silicified fault zones which are more resistant to erosion than adjacent rocks and produce linear topographic features such as ridges or parapets.

V. Landforms produced by mass wasting, landsliding, and soil movement.
- A. Small scale features associated with soil movement such as slump corrugations and patterned soil.
- B. Larger scale topographic features resulting from downward gravitative dislocation of consolidated and unconsolidated materials.
  - 1. Landforms involving slide dislocation of bedrock by slump, block gliding, or rock sliding (Fig. 70).
  - 2. Landforms resulting from slide dislocation of soil and weathered rock (Fig. 70B).
  - 3. Landforms resulting from rapid flowage of unconsolidated materials as in earthflows, debris avalanches, mudflows, and sand or silt flows.

VI. Valleys and canyons.
- A. Valleys and/or canyons cut below a sloping surface on essentially homogeneous rocks. Drainage pattern of main streams and tributaries is dendritic.
- B. Valleys and/or canyons have developed in soft layers in tilted layered rocks. Drainage pattern reveals locations of soft layers as in trellis or ladder patterns.
- C. Valley or canyon has been eroded by a trunk stream superimposed on rocks of varying resistance but retains its original course.
- D. Valley or canyon has formed by erosion along a fault zone.
- E. Stream-cut valleys or canyons have been modified by glaciation which has steepened walls and removed spurs.
- F. Terraced valleys or canyons.
- G. Hanging valleys.

TABLE XVI (continued)

VII. Landforms associated with glaciation.
- A. Valley or alpine glaciation.
  1. Erosional landforms including U-shaped valleys, glacial cirques, serrate ridges at intersections of cirques, rounded resistant rock masses on floor and walls of valley, and scoured lake basins.
  2. Depositional landforms including lateral moraines, medial moraines, ground moraines, and fluvioglacial outwash plains.
- B. Landforms associated with ice-cap glaciation.
  1. Erosional landforms including scoured and rounded rock masses and excavated depressions in softer rocks. Many lakes and a very irregular drainage pattern characterize landscape.
  2. Depositional landforms including low rounded hills or ridges held up by glacial till or sand. Many lakes and irregular drainage pattern are characteristic of landscape.

VIII. Landforms associated with coastlines and shorelines.
- A. Erosional landforms including terraces and sea-cliffs.
- B. Depositional landforms including beaches, bars, and spits.

IX. Extensive surfaces of flat or low relief.
- A. Erosional surfaces.
  1. Rock-floored plains developed by erosional stripping of flat-lying layered rocks. Alluvium and/or soil thin or absent.
  2. Pediments. Gently sloping surfaces cut in rock by sheetwash at base of elevated ground, especially in arid and semiarid regions. Surface cover thin or absent.
- B. Depositional surfaces.
  1. Alluvial plains, flood plains, deltaic plains, and valley-fill plains.
  2. Lava plains, formed by lateral flow of liquid lava, especially basalt or by deposition of extensive horizontal layers of pyroclastic materials.
  3. Alluvial fans.
- C. Composite plains formed by a combination of erosion and alluvial deposition.

X. Miscellaneous landforms.
- A. Karsts formed by collapse of solution caverns in limestone.
- B. Inselbergs (island mountains) standing alone and above surrounding land surface. Generally consist of massive crystalline rocks.
- C. Sand dunes and wind-blown loess deposits.

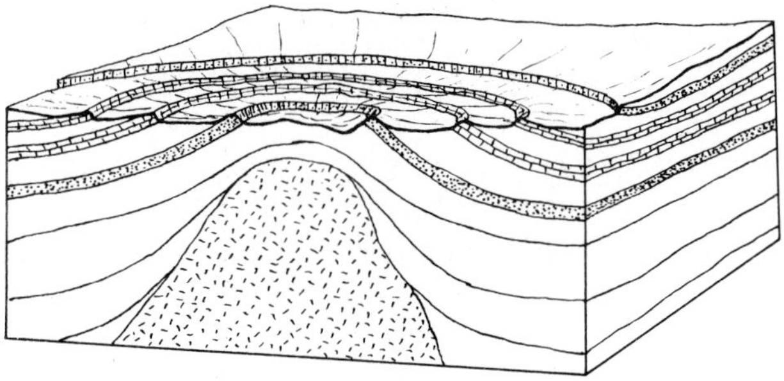

Fig. 64. Eroded dome in layered rocks above stock-like instrusion.

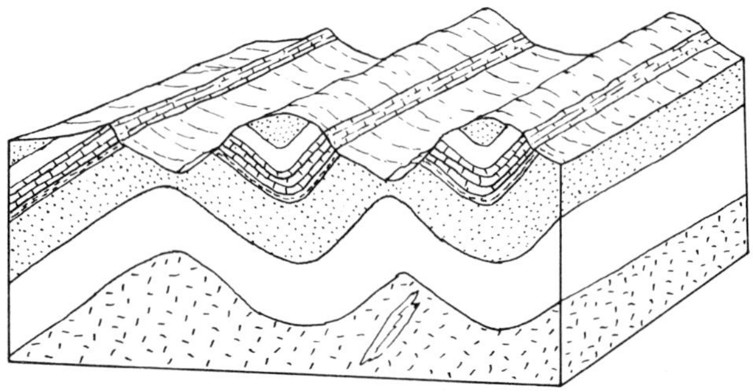

Fig. 65. Parallel valleys eroded in folded sedimentary rocks. Stream valleys have developed in anticlines causing a reversal of original topography. Characteristic of Appalachian Mountains.

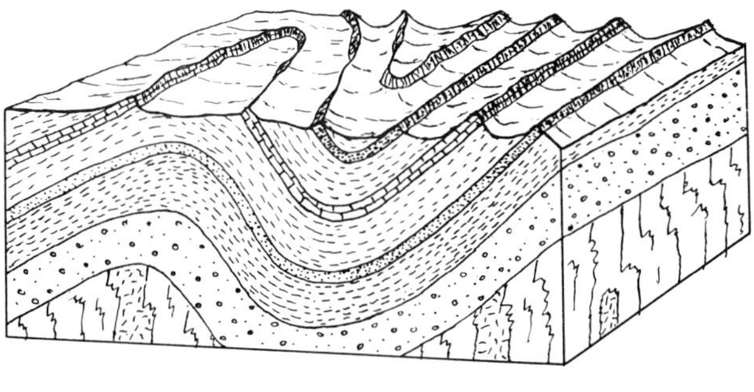

Fig. 66. Landscape formed by erosion of plunging anticline and syncline in sedimentary rocks.

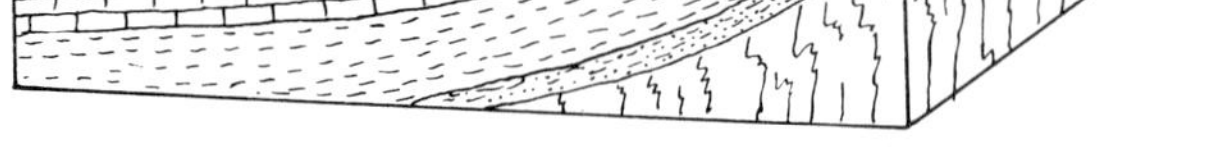

Fig. 67. Hogbacks and mesas formed in sedimentary rocks by differential erosion.

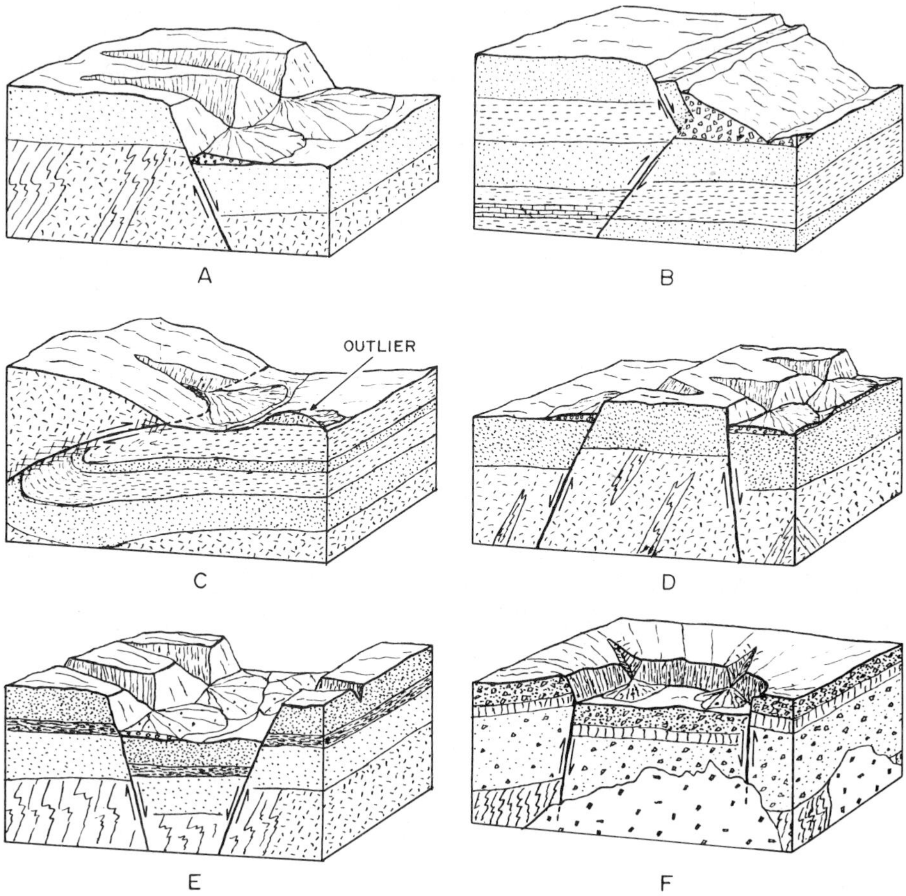

Fig. 68. Several topographic features developed along faults. A. Normal fault has developed a fault scarp. Note triangular facets resulting from stream erosion. Characteristic of geologically recent faults. B. Reverse fault. Scarp developed by slump suggests a normal fault dipping in opposite direction. C. Fault-line scarp developed by erosion along a flat thrust fault. Note outlier which has been isolated by erosion. D. Horst with stream valleys cut in uplifted block. E. Graben with stream deposits covering floor of trench between faults. F. Caldera or volcanic sink in volcanic rocks produced by subsidence into a magma reservoir along a circular fault system.

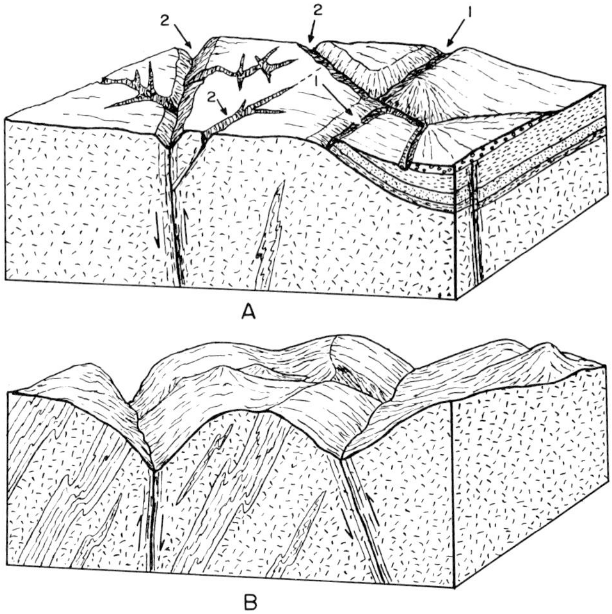

Fig. 69. Linear stream valleys cut into fault zones and soft sedimentary rocks. A. Surface predating downward cutting has low relief. *1* = strike valley in soft sedimentary layer; *2* = valleys following fault zones. B. Deep valleys following fault zones in a mountainous region of crystalline rocks.

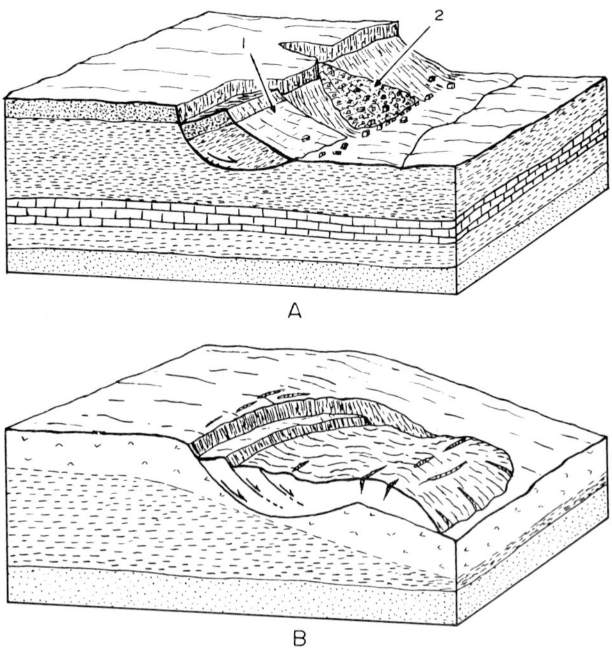

Fig. 70. Some topographic features produced by gravitative dislocation of slopes. A. A slump block (*1*) has formed by dislocation of bedrock, and (*2*) a talus cone has resulted from accumulation of material from a succession of rock falls. B. A landslide has developed in soil and weathered shale. Note scarps and cracks.

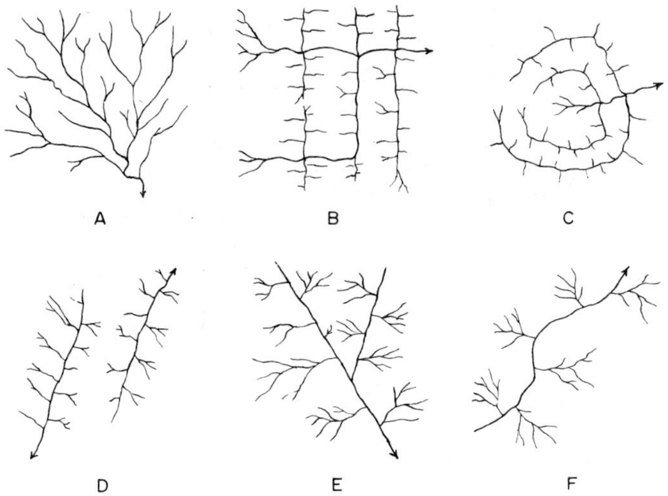

Fig. 71. Some typical drainage patterns. A. Dendritic drainage pattern developed by erosion of rocks of approximately equal resistance to erosion. B. Trellis (ladder) drainage pattern in folded (or faulted) layered rocks. Streams are present between parallel ridges or hogbacks. C. Annular (pericentral) drainage developed in a dome in layered rocks. D. Drainage pattern associated with plunging folds in layered rocks. E. Drainage pattern in a faulted area. Streams are essentially straight in valleys cut into faults. F. A drainage pattern developed along a flat thrust fault.

GEOLOGIC MAPS AND CROSS-SECTIONS OF TUNNEL SITES

Introduction

Tunnels require prior estimation of geologic conditions that will be encountered so that plans may be made to anticipate these conditions in the design of the tunnel and preparation of cost estimates. Simply stated, a tunnel is primarily a geologic problem, and all engineering aspects of tunneling must depend ultimately on the kinds and severity of the geologic conditions that will be met during the excavation of the tunnel. A poor or inadequate investigation of the geology of a tunnel site is an invitation to disaster, as attested to by numerous examples, where unexpected geologic conditions caused construction costs to soar far beyond expectations, and, in many instances, resulted in loss of human life and equipment.

Geological investigations of proposed tunnel locations commonly are conducted in three stages. In the initial stage a geological reconnaissance is made using available maps, aerial photography and, possibly, radar or infrared scanning imagery. The purpose of the reconnaissance is simple: to obtain a first, gross impression of the geologic conditions and to plan subsequent investigations. The second stage requires more detailed investigation and is geared to a determination of the feasibility of a particular location. At this stage consideration is given to alternate tunnel alignments, based on comparison of geologic conditions in the general location of the proposed tunnel route. Good practice may dictate core drilling and/or geophysical studies, and collection and laboratory examination of rock samples. Once a tunnel site is selected, investigations enter the third stage where special additional studies are made to assist in the final design and estimation of the costs of the tunnel.

There is a trend, much to be resisted, of preparing design stage geologic maps and cross-sections primarily from aerial photographs or with remote sensing imagery with little or no effort given to on-the-ground geologic mapping and subsurface exploration. Part of the difficulty stems from the geologists themselves, many of whom find it much easier quickly to answer the demands of the designing engineers by providing apparently knowledgeable interpretations of photographs than to make detailed, time-consuming field investigations. In spite of the elegance and high degree of refinement of modern techniques for photographing or sensing ground conditions from the air, there is no substitute for ground investigations, including geophysical and geological field work, core drilling and testing of bedrock beneath covered areas, and laboratory study of rock samples collected in the field.

The engineer's primary concern is with the geological factors that must be taken into account in design and preparation of cost estimates, and, in general, he is not concerned with the names of rocks, details of rock fabric, or geological theories or abstractions, but, rather, with the practical aspects of tunneling as related to the rate of progress, costs, and safe construction of a tunnel. It is a mistake for a geologist associated with engineering planning and design to extend guesswork beyond reasonable limits, but it is equally a mistake for the engineer to ask the geologist to provide interpretations of bedrock geology based on anything less than a very careful and thorough geological investigation.

A primary purpose in making an accurate and complete geologic map of a tunnel location is the construction of a longitudinal section that will indicate to the extent possible the kinds of geological conditions that will be encountered as the tunnel is driven. Core drilling and geophysical measurements aid in the preparation of the section. The amount of geological information available for use in construction of the section differs widely, depending on surface conditions and the completeness of the geologic investigation. The accuracy of projection of surface geologic features to tunnel level is at a maximum in regions of sedimentary rocks where surface exposures are abundant, and is at a minimum in regions of complex geology where alluvium and soil conceal most or all of the bedrock exposures.

Geologic maps are prepared with an appropriate scale, and rock types and named formations are indicated in color or by patterns. Commonly, letter symbols are used on the map to indicate rock types, in addition to colors or patterns. Symbols

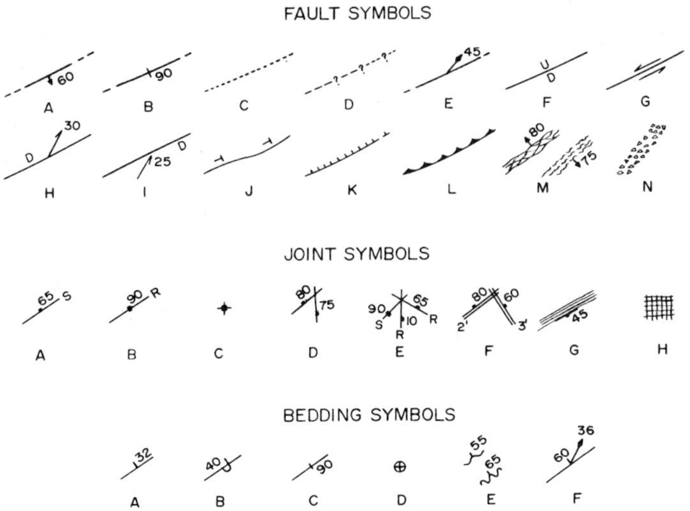

Fig. 72. Symbols used on geologic maps (mainly after conventions used by the U.S. Geological Survey).

Fault symbols:

A. Strike and dip of fault (dashed where approximately located).
B. Strike and dip of vertical fault.
C. Concealed fault.
D. Doubtful or probable fault.
E. Fault showing bearing and plunge of grooves, striations, or slickensides.
F. High angle fault showing upthrown side (*U*) and downthrown side (*D*).
G. Fault, showing directions of relative movement.
H. Normal fault, showing bearing and plunge of relative movement of downthrown block (*D*).
I. Reverse fault, showing bearing and plunge of relative movement of downthrown block.
J. Thrust or low angle reverse fault. Upper plate designated by horizontal T.
K. Normal fault, hachures on downthrown side.
L. Thrust or reverse fault, saw-teeth on side of upper plate.
M. Fault zone or shear zone.
N. Fault breccia.

Joint symbols:

A. Strike and dip of smooth (*S*) joint.
B. Strike of rough (*R*) vertical joint.
C. Horizontal joint.
D. Intersecting joints.
E. Intersecting joints, smooth (*S*) and rough (*R*).
F. Intersecting joint sets showing average spacing in feet of joints in each set.
G. Strike and dip of sheeted zone.
H. Blocky jointing.

Bedding symbols (sedimentary rocks):

A. Strike and dip of beds.
B. Strike and dip of overturned beds.
C. Vertical beds.
D. Horizontal beds.
E. General strike and dip of crenulated, crumpled, plicated, or undulating beds.
F. Strike and dip of beds and direction and plunge of bedding-plane slickensides.

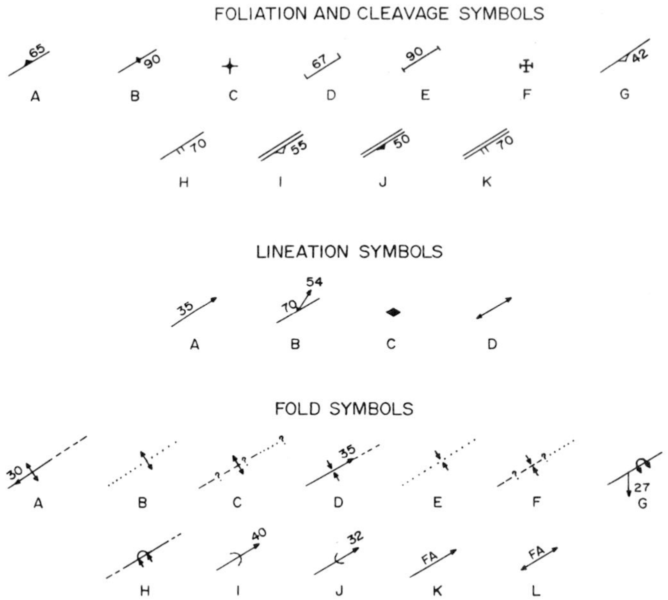

Fig. 72 (continued)

Foliation and cleavage symbols (metamorphic rocks):
A. Strike and dip of foliation.
B. Strike of vertical foliation.
C. Horizontal foliation.
D. Strike and dip of cleavage.
E. Vertical cleavage.
F. Horizontal cleavage.
G–K. Special purpose symbols to be used as required.

Lineation symbols:
A. Bearing and plunge of lineation.
B. Strike and dip of foliation and bearing and plunge.
C. Vertical lineation.
D. Horizontal lineation.

Fold symbols:
Note: If crest line is mapped, i.e. the highest portion of an anticline or the lowest portion of a syncline, an explanation should be given with the map.
A. Anticline showing trace of axial plane and bearing and plunge of axis.
B. Concealed anticline.
C. Doubtful or probable anticline.
D. Syncline showing trace of axial plane and bearing and plunge of axis.
E. Concealed syncline.
F. Doubtful or probable syncline.
G. Overturned anticline showing trace of axial plane and bearing and plunge of axis.
H. Overturned syncline showing direction of dip of limbs.
I. Plunge of minor anticline.
J. Plunge of minor syncline.
K. Plunge of fold axis.
L. Horizontal fold axis.

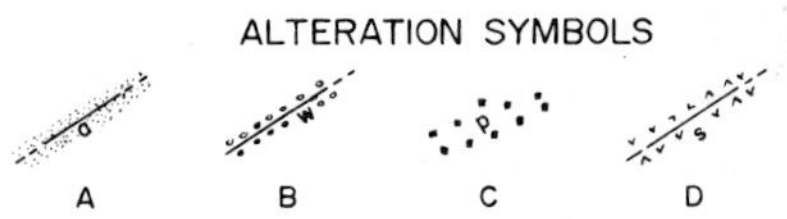

Fig. 72 (continued)

Alteration symbols:
A. Argillic alteration along a fault.
B. Weathering alteration along a fault.
C. Pyritization.
D. Silicification along a fault.

for structural and other features are fairly well standardized and are indicated in Fig. 72.

Kinds of surface geologic maps

Geologic maps are made for a variety of purposes by utilizing a broad spectrum of techniques. Three kinds of geologic maps, their method of preparation, and their advantages are described in subsequent paragraphs.

(1) Geologic maps based only on aerial photographs and/or on radar or thermal imagery

Preparation of such maps relies heavily on subjective interpretation of the geological significance of the distribution and shapes of visible rock exposures, distribution and shapes of deposits of alluvium, characteristics of landforms, distribution and character of vegetation, or thermal effects produced by warm or cold surface or groundwater bodies or bedrocks of different thermal properties. The techniques of mapping are especially useful in regions where bedrock exposures are at a minimum or in regions of layered rocks where bedrock exposures are abundant.

In regions of extensive cover the tendency of the observer is to interpret every linear element on a photograph or image as a bedrock fault, and a typical map may indicate many more faults than actually exist. Such mapping is not satisfactory in the detection of important features of bedrock such as close jointing or alteration and detailed lithology, and without careful and thorough field checking, is of limited value.

(2) Interpretive bedrock geologic maps based on ground observation and mapping on contoured topographic base maps

Commonly aerial photographs and/or radar or thermal imagery are used to locate and identify features of interest. The kind of map that results depends to a large extent on the purpose of the map, the time expended in its preparation, and whether emphasis is placed on surficial geology or bedrock geology.

Bedrock maps are made by field location and examination in the field of all bedrock exposures and, where bedrock is covered by soil or products of mass wasting, by "float" mapping which takes account of the distribution of products of weathering and slope reduction. In regions of extensive cover bedrock maps prepared in this manner are highly subjective and in many instances represent only intelligent guesswork by a trained field investigator as to bedrock conditions in covered areas. Frequently bedrock exposures form only in highly competent, resistant rocks, and soft, incompetent rocks, fault zones, and altered rocks, features of special concern in tunneling, are covered. Appropriate assessment of bedrock geology requires core drilling and/or geophysical measurements.

Because glacial and stream deposits have been transported to the site of deposition by moving ice or water, float mapping of these deposits is ineffective, and most geologists do not attempt to project bedrock geology under such deposits, if they are widespread.

The use of highly interpretive bedrock maps in detailed planning and design of tunnels, especially in regions of complex geology and extensive surface cover, is at best hazardous, and such maps should be used only in the identification of regional fabric and rock types. An exception should be noted for maps prepared in mining districts where extensive surface and underground workings enable minute examination of bedrock geology and safe projection to the surface.

(3) Surface exposure maps

Maps which are prepared in the field with the aid of topographic base maps, aerial photographs, and radar or thermal imagery, and show the locations and detailed geology of actual bedrock exposures, and the distribution and characteristics of surface cover are among the most useful maps in geological investigations of engineering projects. Such maps may show interpretations of bedrock geology under covered areas, but the user of the map has a clear concept of the extent and nature of the data used in making the interpretations, and, if he so desires, he may make his own interpretations and projections. A particular advantage of an exposure map is that it makes a clear distinction between what is known and what must be inferred or guessed at and provides a basis for intelligently planned additional exploration of subsurface conditions by core drilling and/or geophysical investigations.

Starting with an exposure map, the geologist may make a second map which shows his concept of the bedrock geology that might be observed if all surface cover were removed, but this map should be clearly identified for what it is, and its limitations specified. Indeed, it is the function of the geologist to prepare such a map, for he is the only person with the necessary training, experience, and intuitive capacity.

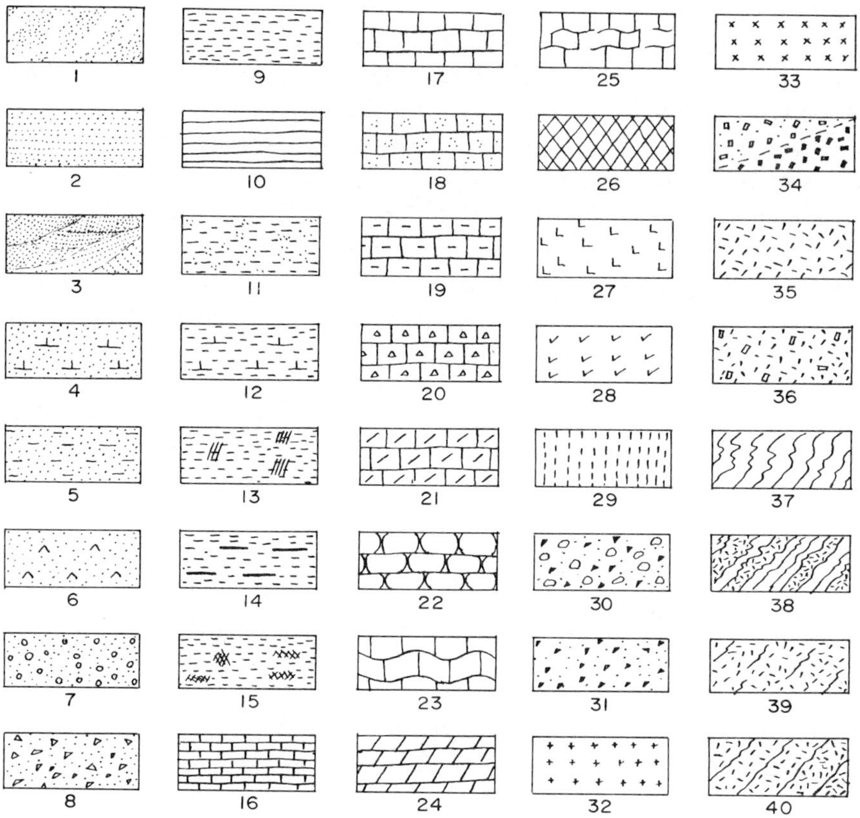

Fig. 73. Commonly used lithologic symbols.

1. Massive sandstone
2. Bedded sandstone
3. Cross-bedded sandstone
4. Calcareous sandstone
5. Shaly sandstone
6. Quartzite sandstone
7. Conglomerate
8. Sedimentary breccia
9. Shale
10. Mudstone or claystone
11. Sandy shale
12. Calcareous shale
13. Fireclay
14. Carbonaceous shale
15. Gypsiferous shale
16. Thin-bedded limestone
17. Massive limestone
18. Sandy limestone
19. Shaly limestone
20. Cherty limestone
21. Dolomitic limestone
22. Concretionary limestone
23. Chalk
24. Dolomite
25. Marl
26. Gypsum or anhydrite
27. Rock salt
28. Tuff
29. Bentonite
30. Agglomerate (igneous)
31. Pyroclastic material
32. Lava
33. Intrusive igneous rock
34. Porphyry
35. Granite (or phaneritic igneous rock)
36. Granite porphyry
37. Schist
38. Gneiss or migmatite
39. Gneissoid granite
40. Granite gneiss

Geologic cross-sections

Geologic cross-sections are projections of surface geology to depth and in tunneling investigations usually are constructed in a vertical plane including the tunnel alignment. Geological projections of necessity are based entirely on surface geologic mapping, supplemented by geophysical data, information derived from core drilling, or, in some areas, in data derived from mine workings.

As in geologic maps, geologic sections are drawn with varying degrees of certainty. A section in an area of extensive surface cover and complex bedrock geology generally is boldly hypothetical, and actual conditions encountered in a tunnel may differ radically from uncertain, highly-subjective projections by the geologist based on meager surface data. It is questionable whether such sections should appear in reports of geological investigations of engineering projects, because they may give an entirely erroneous impression of the geologic conditions of concern to the designing engineer and the contractor who drives the tunnel. On the other hand, the purpose of surface geological investigations is predetermination of the geological conditions at depth to the extent possible, and if some attempt is not made to estimate the extent and kinds, but not necessarily the exact locations of geologic conditions at tunnel level, the geological investigation serves no useful purpose.

In preparation of geologic sections for use in engineering projects, it is incumbent on the geologist to state explicitly or to indicate by means of symbols and text explanation the presumed accuracy of his projections. The wise engineer will understand that allowances should be made for the unexpected.

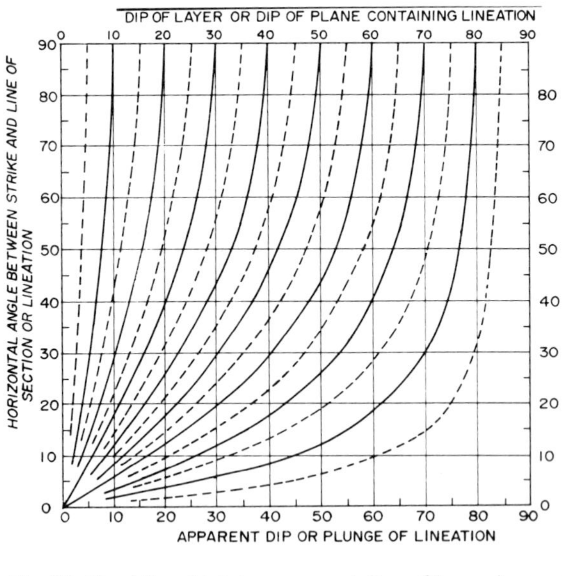

Fig. 74. Chart for obtaining apparent dips of layers in cross-sections or plunges of lineations.

Fig. 73 indicates symbols commonly used in cross-sections and columnar sections for indicating lithologic types. Fig. 74 is a chart useful in obtaining apparent dips in geologic sections when the line of section is not at right angles to the strike of layered rocks or other planar features such as fractures. Fig. 74 also may be used to determine the angle of plunge of linear features in planes when the strike of the plane (such as a foliation plane) and the horizontal direction of the linear feature are known.

Columnar sections

In areas of sedimentary rocks one of the first tasks of the geologist is the identification of the lithology of the various sedimentary layers and the measure-

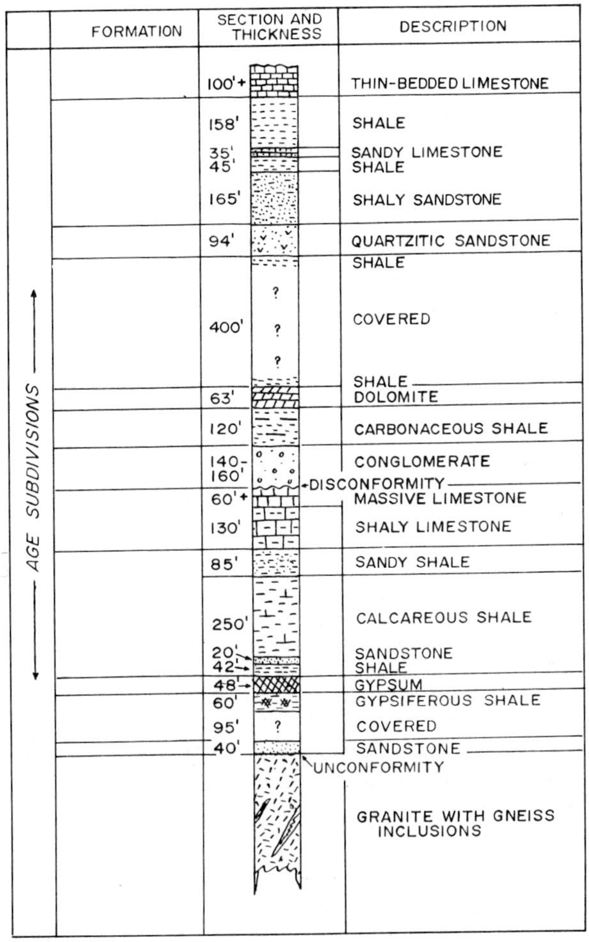

Fig. 75. Elements of a columnar section. Age subdivisions and formation names are omitted.

ment of their thicknesses. In regions of extensive cover and complex folded and faulted rocks the task may be difficult or practically impossible, but, nevertheless, every effort should be made to establish the sedimentary sequence. If economic or engineering factors indicate need for a knowledge of the details of a sedimentary sequence, core drilling in selected locations may be justified.

Sedimentary sequences commonly contain "marker beds" which, because of resistance to erosion and distinctive lithology, are easily recognized in the field and are especially useful in working out details of folding and faulting in the bedrock and in construction of geologic cross-sections from surface maps.

Very few areas permit measurement of thicknesses and determination of lithology along a single line at right angles to the strike of the layers. Accordingly, it is usual practice to construct columnar sections by taking measurements in a variety of locations and to prepare a composite section based on these various measurements. In areas where beds are highly lenticular estimates should be made of the range of thicknesses for each sedimentary unit.

Fig. 75 indicates some of the kinds of information that may be included in a columnar section (also called a "stratigraphic section").

Geologic and geophysical data from field and laboratory investigations

In geological investigation of a proposed tunnel location and in preparation of geological maps and cross-sections every effort should be made to gather data concerning gross geological features as well as all small-scale features that might have an effect on the progress of the tunneling operation. A check list of observations that should be made, with explanatory text or elaborations is given in Table XVII.

TABLE XVII

Check list for data gathering in geological investigation of proposed tunnel sites

Data should be presented in maps and cross-sections, columnar sections, and in tables, graphs, and diagrams, together with narrative, descriptions, and interpretations in the body of a report accompanying the maps and sections.

I. Surficial deposits.
Determine characteristics and origin of surficial deposits and locate various types on geologic map. Examine particularly surficial deposits of local origin to determine lithology and to estimate competence of source bedrock.

II. Geomorphology.
Examine landforms and classify as to type and origin. In particular, pay attention to shapes and distribution as they reflect bedrock lithology, subsurface structure, and alteration. If possible, correlate steepness of slopes with competency of bedrock.

TABLE XVII (continued)

III. Identify various lithologic units and map exposures. Make careful notes of megascopic mineralogy and primary fabric of each unit. Collect samples for petrographic examination in the laboratory and, if appropriate, for laboratory tests of engineering properties.

a. For igneous rocks, note and map planar and linear original fabrics, if they are present, particularly flow layering and flow lineation.

b. For sedimentary rocks make every effort to measure and prepare a stratigraphic section (columnar section) showing detailed thicknesses and lithologies of individual layers and unconformities and disconformities. If necessary, and the need for detailed information is critical, use a core drill to obtain samples.

c. In metamorphic rock exposures note lithology and primary fabrics and pay particular attention to kinds and attitudes of foliation, cleavage, and lineation. In complex metamorphic terranes look for layers of distinctive lithology and fabric that might serve to define structural relationships. Note and map major and minor folds.

IV. Identify all features of secondary anisotropism including joints, faults, and folds, and locate on geologic map with appropriate symbols and annotations.

a. For faults attempt to determine dip, strike, and direction and amount of relative movement of adjacent blocks. Pay particular attention to width and characteristics of disturbed rock associated with faults.

b. Because of their prime importance to the tunnel engineer concerned with an evaluation of tunnel excavation techniques and with an estimation of the quantity and suitability of various kinds of supports, joint studies in the field should be made very thoroughly and carefully. Joints of concern are those which can be expected to exist in the body of the rock at depth, and care should be taken to eliminate from consideration joints resulting from surface weathering, glaciation, or mass slope failure.

Attempt to identify distinguishable joint systems and note and plot attitudes. Pay attention to and record spacings of joints in each system, lateral and/or vertical extent of jointing in each system, roughness or smoothness of the joints, and relative persistence of intersecting joint systems.

Data should be incorporated in equal-area plots or some other type of graphical representation. Computerization often is of assistance in analyzing joint data.

c. In folded rocks note type of folds and pay particular attention to movement of adjacent layers, development of fractures in crests, troughs, or limbs, and behavior of incompetent layers between more competent layers.

V. Attempt an evaluation of ground-water conditions. Locate, map, and estimate quantity of water issuing from springs. Make special note of thermal springs which might indicate high temperatures at depth. Evaluate, to the extent possible, secondary permeability associated with fractures and dissolution of limestone.

VI. Estimate the possibilities of underground gas. Pay particular attention to the likelihood that explosive methane gas may be present in or derived from carbonaceous shales or limestones. Carbon dioxide commonly is associated with springs, especially with igneous activity, and may be present in large amounts at depth.

VII. Estimate rock temperatures at tunnel level. High temperatures commonly are locally present in regions of recent orogenic or igneous activity, and may be indicated by thermal springs at the surface.

VIII. Locate areas where geophysical investigations and core drilling will contribute to a knowledge of the bedrock geology, and correlate the results with other data.

GEOPHYSICAL INVESTIGATIONS

Introduction

Geophysical investigations give valuable assistance in the determination of subsurface geologic conditions and are of particular use especially where mapping of surface geology yields inconclusive data concerning bedrock geologic conditions. Geophysical exploration is essentially a method of measurement of physical quantities, and depends to a large extent on identification of measurable differences in the physical quantities associated with differences in structure or lithology in subsurface assemblages of rocks.

The major techniques of geophysical investigation in a general classification break down into categories including (1) measurement of variations of the earth's magnetic and gravitational fields, (2) measurement of differences in reaction of earth materials to artificially induced force fields produced by electrical or seismic impulses introduced into the rocks at or near the earth's surface, and (3) measurement of electrical currents spontaneously generated by geologic bodies. Several kinds of geophysical investigations result in the identification of one or more "anomalies", that is, departures from the average responses obtained by measurements. It is the task of the geologist in cooperation with the geophysicist to interpret the significance of these anomalies.

Within the limited scope of the present treatment, no attempt can be made to give more than brief mention to the extensive theory and practice of geophysics.

TABLE XVIII

Outline summary of geophysical methods for subsurface exploration

I. Seismic methods utilizing artificially induced earth tremors, usually created by explosives or weight impacts.
   A. Refraction method.
   B. Reflection method.

II. Gravity methods measuring variations in the earth's gravitational field as related to subsurface geological structure.

III. Magnetic methods measuring contrasts or variations in magnetic susceptibilities of rocks.

IV. Electrical methods.
   A. Methods measuring natural potentials.
   B. Methods measuring potential drop in current transmitted between electrodes.
   C. Methods measuring distortions in natural or induced electrical and magnetic fields.

V. Methods based on measurement of radioactivity of rocks.

Instead attention will be given to broad categories of investigation with brief descriptions of a few commonly used applications.

For detailed discussions of geophysical theory and practice the reader is referred to the extensive literature but especially to treatments by Dobrin (1960) and Grant and West (1965).

Table XVIII summarizes the categories of geophysical investigations that have proved to be of value in engineering investigations.

Seismic methods

Several kinds of elastic waves are generated by seismic disturbances whether natural or induced. In seismic prospecting longitudinal (compressional) waves have been used almost exclusively because they are usually the first to arrive at detecting instruments, although some attempts are being made to utilize slower transverse waves (also called *shear waves*) in the study of rock structures.

The speed of longitudinal waves, V_L depends on the elastic constants of rocks according to the relationship in eq. 7–1:

$$V_L = \sqrt{\frac{E}{\rho} \frac{1-\sigma}{(1-2\sigma)(1+\sigma)}} \tag{7-1}$$

In this equation E is Young's modulus (modulus of elasticity), ρ is density, and σ is Poisson's ratio, usually with a numerical value near 0.25 but not in excess of 0.50. It is apparent that the velocity depends to a great extent on the value of Young's modulus, E.

The refraction method is widely employed in engineering investigations. Until recently, the method was used principally in measuring depths of overburden in tunnel portal areas, but there is increasing activity in its use in locating faults and in estimation of the quality of bedrock. A discussion of the use of seismic methods in bore holes is presented later in this chapter under the section on exploratory core drilling.

Seismic methods depend on the fact that there are differences in the velocities of transmission of seismic impulses among rocks (Table XIX). Seismic velocities depend on several factors including chemical composition, degree of lithification, mineral composition, fabric, and fluid saturation. In general, velocities are least in unconsolidated deposits and at a maximum in crystalline rocks. Secondary factors tending to decrease velocities include weathering, extensive fracturing, an increase in porosity, and hydrothermal alteration, all of which tend to reduce density.

Fig. 76 shows some paths followed by artificially generated seismic impulses. Fig. 77 illustrates the basis for determining depths of layers in a simple situation,

TABLE XIX

Some velocities of seismic longitudinal (compressional) waves in natural materials (From F. Press, 1966)

	Velocity (ft./sec)	*Velocity* (km/sec)
Alluvium	1,640– 6,600	0.5–2.1
Clay	3,600– 8,200	1.1–2.5
Loess	980– 1,200	0.3–0.6
Sand	650– 6,600	0.2–2.0
Glacial till	1,300– 5,600	0.4–1.7
Granite, quartz monzonite, granodiorite	15,100–19,700	4.6–6.0
Gabbro, diabase, basalt	16,400–22,000	5.0–6.7
Sandstone, shale	4,600–14,800	1.4–4.5
Limestone		
soft	5,600–13,800	1.7–4.2
hard	9,200–21,000	2.8–6.4
crystalline	18,700–21,000	5.7–6.4
Anhydrite, gypsum, salt	11,500–18,000	3.5–5.5
Slate	11,500–14,400	3.5–4.4
Schist and gneiss	11,500–24,600	3.5–7.5

and suggests the kinds of considerations that are useful in determining the geometry of subsurface geologic structures.

In Fig. 77 a seismic impulse at a shot point is transmitted along various paths to geophones, and a time–distance graph is plotted from the times of first arrival of impulses at each geophone. The velocities of each of the layers is obtained by

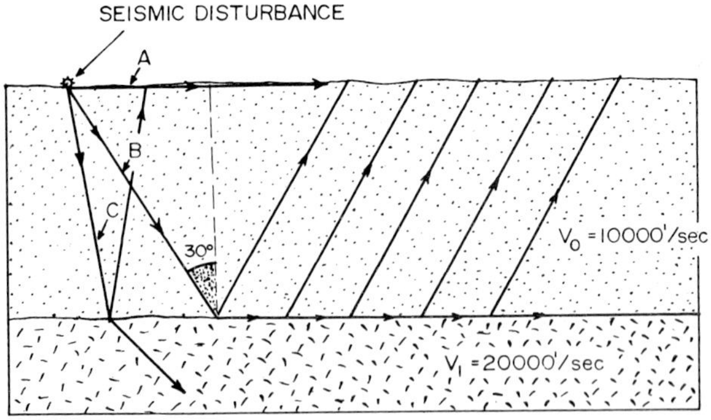

Fig. 76. Some of the paths followed by energy from an artificially generated seismic impulse. *A* = energy travels parallel to earth's surface; *B* = energy is refracted at critical angle, travels in medium of velocity V_1, and provides impulses which return to surface; *C* = energy is partly refracted into medium of velocity V_1 and is partly reflected back to the surface.

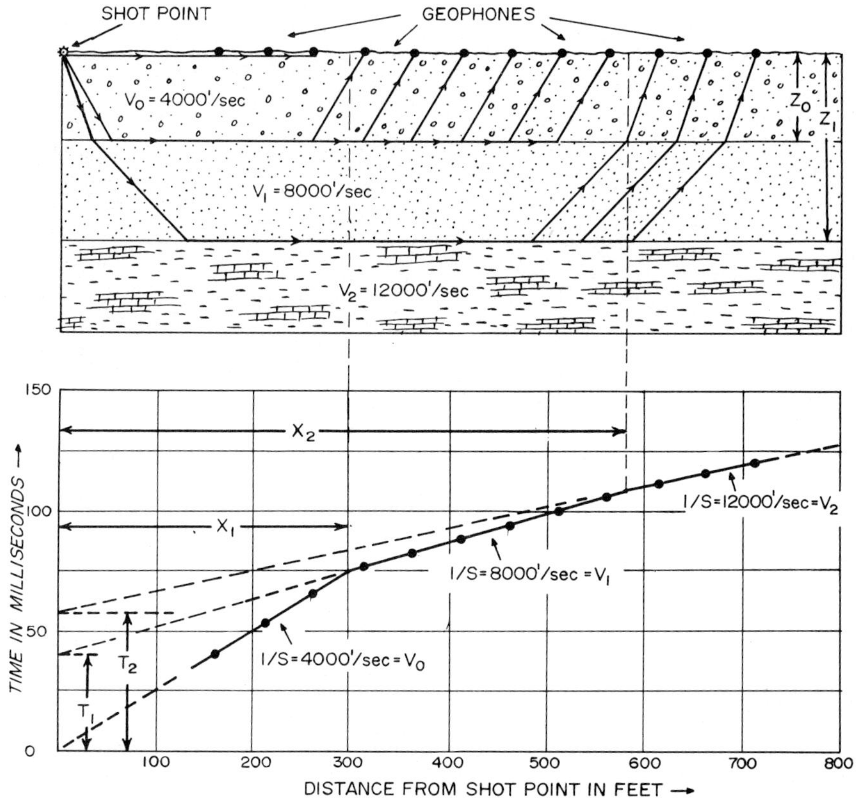

Fig. 77. Paths traveled by seismic impulse which provide first arrival at geophones in a seismometer, and time–distance curves obtained from measurements of time of arrival and distances of geophones from shot point.

taking the reciprocal of the slopes of each segment of the time–distance curve. Inspection of the time–distance curves yields intercept times T_1 and T_2, obtained by projection to the time coordinate of the segments of the time–distance lines, and *critical distances,* X_1 and X_2, which are identified by intersections of the segments of the time–distance curves.

The depths to the layers Z_0 and Z_1 can be calculated by the following formulas, which are most expeditiously solved by programming into a computer:

$$Z_0 = \frac{X_1}{2}\sqrt{\frac{V_1 - V_0}{V_1 + V_0}} \tag{7-2}$$

or:

$$Z_0 = \frac{T_1 V_0 V_1}{2\sqrt{V_1^2 - V_0^2}} \tag{7-3}$$

and:

$$Z_1 = \frac{X_2}{2}\sqrt{\frac{V_2 - V_1}{V_2 + V_1}} + Z_0\left[\frac{V_2(V_1^2 - V_0^2)^{1/2} - V_1(V_2^2 - V_0^2)^{1/2}}{V_0(V_2^2 - V_1^2)^{1/2}}\right] \tag{7-4}$$

or:

$$Z_1 = V_1\left[\frac{T_2 - (2Z_0/V_0)[(V_2^2 - V_0^2)/V_2^2]^{1/2}}{2[(V_2^2 - V_1^2)/V_2^2]^{1/2}}\right] \tag{7-5}$$

Gravity methods

Gravity exploration of subsurface geologic features detects and measures lateral variations in the earth's gravitative field that are associated with near-surface changes in density. Commonly, a very sensitive instrument called a *gravimeter* is used to identify gravity anomalies above bodies of rock with smaller or greater densities than adjacent rocks. Before anomalies can be detected with certainty, a variety of corrections to readings must be made. Generally corrections are made for station elevation with respect to an arbitrarily selected datum plane, latitude, and, in regions of considerable topographic relief, for the influence of nearby topography. A possible use of gravity measurements in engineering investigations is in the identification of extensively altered zones which may have lower densities than the original rocks.

Magnetic methods

Magnetic methods of subsurface exploration depend on the fact that earth materials in the earth's magnetic field display different *magnetic susceptibilities.* Magnetic susceptibility is the ratio of the degree of magnetization to the intensity of the magnetizing force. Lines of magnetic force in the earth's field tend to concentrate in *ferromagnetic* and *paramagnetic* substances and are dispersed in *diamagnetic* substances such as rock salt and anhydrite. Most rocks are weakly paramagnetic but show appreciable and measurable differences in susceptibility.

Except at the magnetic poles the earth's magnetic field is inclined with respect to the earth's surface. Commonly the magnetic field or its vertical or horizontal components are measured with a dip needle, a sensitive magnetometer containing a carefully balanced magnet, or a magnetometer based on the phenomenon of nuclear magnetic resonance. Magnetometers lend themselves to observations on the ground or from the air, and air-borne magnetometers have widespread use in regional reconnaissance studies.

Electrical methods

Electrical methods of geophysical investigation of subsurface geological structure have been used extensively in investigation of proposed tunnel sites. The methods are most applicable for study of shallow subsurface conditions, at depths not exceeding a few hundred feet. A great variety of techniques are available but only a few will be described here, principally to illustrated the kinds of data that may be obtained.

The *self-potential method* is employed where electric potentials have developed by electrochemical action between minerals and the solutions with which they are in contact. Sulfide ore bodies in the process of oxidation commonly develop strong potential differences by a battery action that results from differential oxidation of one part of the body with respect to another. Measurement of self-potential differences generally has little application in usual engineering studies.

Resistivity methods, on the other hand, frequently yield data assisting greatly in interpretation of subsurface geology, especially in covered areas. Resistivity techniques in general are based on the measurement of *apparent resistivity,* that is resistivity measured in inhomogeneous ground, commonly layered, between two potential electrodes in an electrical field provided by two current electrodes generating either a direct current or a low-frequency alternating current. The units of resistivity in the metric system are the ohm-centimeter or the ohm-meter.

In a general classification materials with resistivities less than 10^{-5} ohm-m are classed as *conductors; semi-conductors* have resistivities between 10^{-5} and 10^{8} ohm-m; and *insulators* have resistivities above 10^{8} ohm-m. Common rock-forming minerals are insulators, and it is evident that natural materials that behave as conductors or semiconductors with low to moderate resistivity owe this property to the amount of water that is present, its salinity, and its distribution. Water samples collected from a variety of porous rocks have resistivities of the order of 0.05–80 ohm-m, values less than the resistivities of dry rocks by many orders of magnitude.

From the above discussion it is apparent that resistivity measurements have a particular value in geological investigations in locating ground-water tables and in identifying buried fractured zones, especially faults, when they are saturated with water-containing electrolytes.

Fig. 78A shows an electrical configuration commonly used in apparent-resistivity measurements in the field. An electrical current is introduced into the ground through two current electrodes, and the potential drop (voltage) is measured between two potential electrodes. The apparent resistivity ρ between the potential electrodes is:

$$\rho = \frac{2\pi V}{I\left(\frac{1}{r_1} - \frac{1}{r_2} - \frac{1}{R_1} - \frac{1}{R_2}\right)} \quad (7\text{-}6)$$

where V is the potential difference across electrodes B and C, I is the current introduced at electrodes A and D, r_1 is the distance from A to B, r_2 is the distance

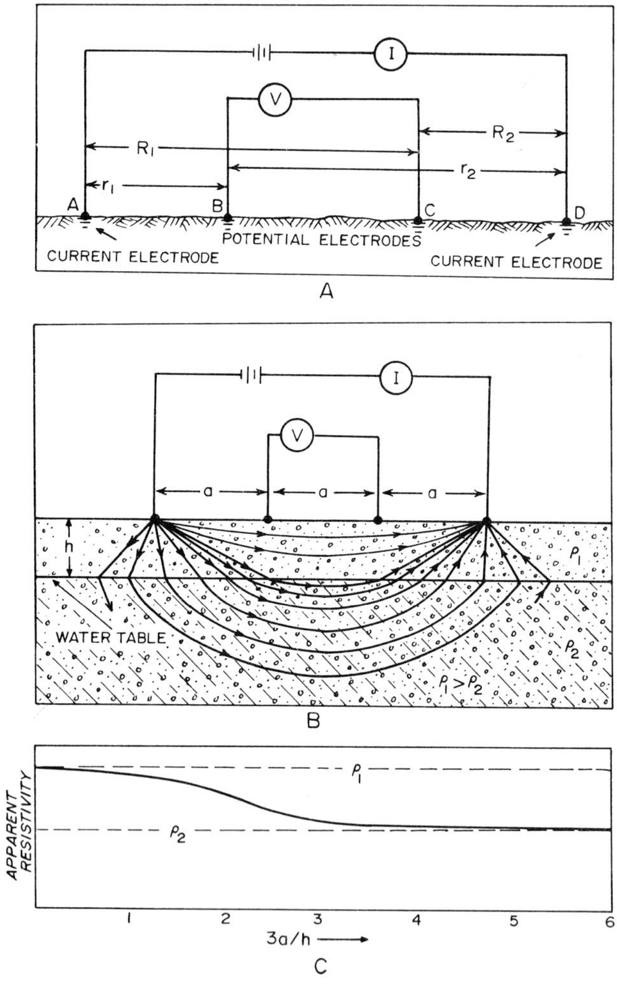

Fig. 78. Electrical configurations utilized in field measurement of apparent resistivity. A. Current electrodes introduce electrical current into subsurface enabling measurement of potential drops between potential electrodes. B. Idealized lines of current flow in a two-layered profile created by ground-water saturation below the water table in alluvium. Configuration of current and potential electrodes is the Wenner configuration. C. Changes in measurements of apparent resistivity as a function of increasing electrode separation (idealized).

from B to D, R_1 is the distance from A to C, and R_2 is the distance from C to D. In the so-called Wenner configuration the potential electrodes are spaced so that $AB = BC = CD$ and eq. 7-6 reduces to:

$$\rho = 2\pi a(V/I) \tag{7-7}$$

where $a = 1/3$ of the current electrode spacing (Fig. 78B).

Fig. 78B diagrammatically shows the flow of electrical current between two current electrodes in alluvium which is saturated with water below the water table. It is assumed that the layer below the water table has a much lower resistance (a much higher conductivity) than the material above the water table. The configuration of the current and potential electrodes is the Wenner configuration in which the interval between the current electrodes is divided into equal segments of length a by the potential electrodes. Fig. 78B suggests that as a is increased so as to become large compared with h, the depth of the ground-water table, the current confined to the layer above the water table becomes negligible. Fig. 78C indicates the change in apparent resistivity as electrode spacing is increased and suggests the basis for estimating the depth to the ground-water table. Although in actual field examples a measured depth to the ground-water table should be correlated with variations in the electrode spacing, it is practicable approximately to estimate the depth of the ground-water table as equal to the electrode spacing, a, when there is

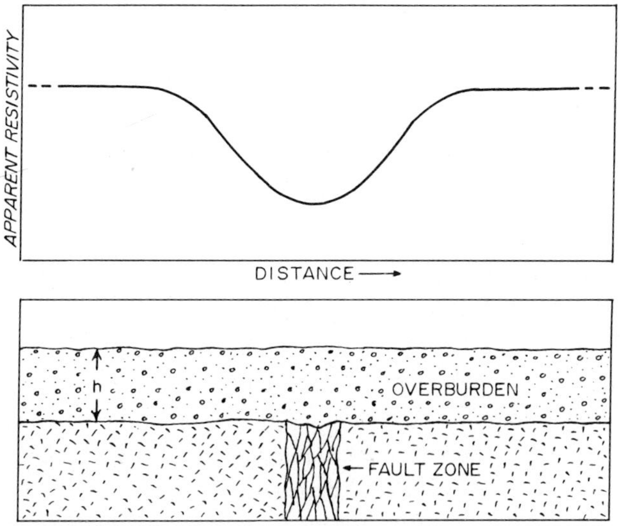

Fig. 79. Idealized apparent resistivity curve obtained by surface measurements with a fixed electrode spacing over a water-saturated fault buried by overburden.

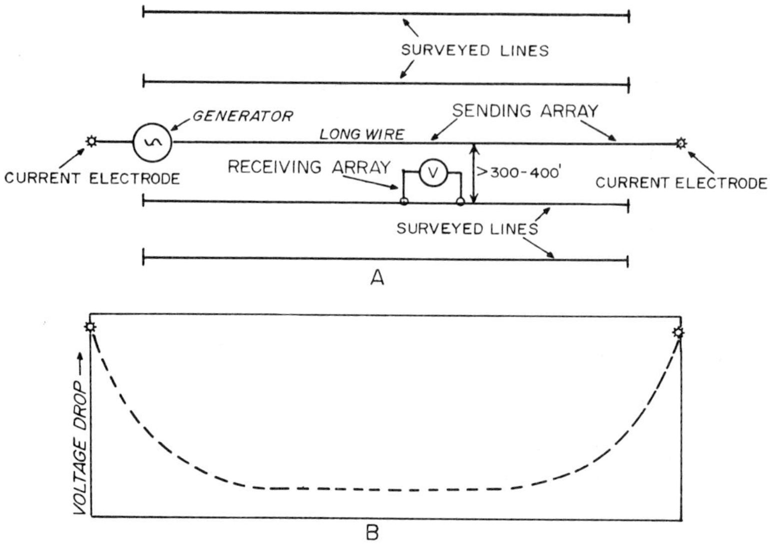

Fig. 80. Elements of the "long-wire" technique for study of resistivities (or conductivities) of subsurface materials. A. Field layout. B. Typical idealized curve for voltage drop between current electrodes.

no appreciable decrease in the apparent resistivity with increase in electrode spacing.

Fig. 79 indicates the kind of apparent-resistivity profile that might be obtained with fixed electrode spacings over a water-saturated fault zone buried by alluvium, and suggests an important application of the resistivity method. It is assumed that the water in the fault zone contains electrolytes and has a considerably smaller resistivity than the adjacent rock or the overburden.

Fig. 80A indicates the elements of another widely used technique for studying variations in the apparent resistivity of materials below the surface. The method, sometimes called the "long-wire method" utilizes a source of alternating current and a long wire extending for hundreds to thousands of feet from one current electrode to the other. A receiving array measures voltages along surveyed lines parallel to the wire, and the results are usually plotted as voltages versus distance. Fig. 80B schematically shows a typical voltage profile in material of uniform resistivity. Notable departures from this kind of profile indicate variations in the resistivity of subsurface materials.

Many techniques have been developed which utilize earth currents and alternating magnetic fields. The currents may be natural currents (telluric currents) or may be induced. Commonly, distortions in natural or induced fields are studied by means of search coils or loops. A widely employed technique is based on the measurement of natural magnetic fields at audio and subaudio frequencies. This technique is the 'AFMAG' (audio-frequency magnetic fields) techniques. With the aid of a coil and earphones (or electronic equipment) measurements are made of the tilts from the horizontal of the plane of polarization of oscillating magnetic fields distorted by subsurface conducting bodies.

Geophysical methods based on *measurements of radioactivity* are directed primarily toward discovery of economic deposits and have little application to engineering studies.

REMOTE SENSING INCLUDING AERIAL PHOTOGRAPHY

Aerial photography and radar and thermal sensing from the air have arrived at an advanced state of sophistication, in part because of the efforts of the armed forces to find methods for location and interpretation of distant ground and submarine features of military concern. Engineers and geologists have put the advancing technology to good use, and aerial photography and remote sensing are now considered as routinely essential to investigation of a great variety of geologic problems. Preparation of contoured topographic base maps from black and white and color aerial photographs by photogrammetric methods has long been accepted as the most economical and accurate method for mapping the configuration of ground terranes.

Interpretation of subsurface features from aerial photographs or imagery commonly is based on conspicuous to very subtle variations in the appearance of the terrain from one location to another depending on differences in vegetation, soils, soil colors, rock colors, and the like. During the Second World War intelligence officers discovered that color-blind individuals could see things in color aerial photographs that were not apparent to persons with normal color vision, and the advantages of taking photographs with a variety of color filters or of viewing normal color photographs through a series of filters became apparent.

Largely because of the needs of the military and geo-scientists, rapid advances have been made in the use of television, radar, infrared photography, and thermal sensing equipment in the mapping of terrain, in addition to the conventional methods of aerial photography.

Aerial photography and remote sensing provide the geologist with powerful tools for reconnaissance investigation of the geology of an area of interest and enable him efficiently to plan the program of ground investigation by standard geological and geophysical methods, which is necessary to any adequate study. Much time is saved by photographs and/or imagery which pinpoint rock exposures, reveal subsurface folded structures with vague or inconclusive surface manifestations on the ground, and locate lineaments along zones of fracture. Intelligent interpretation of the origin and significance of land forms, distribution of vegetation, stream patterns, and distribution and origin of soils and overburden are expedited by photographs and images. However, it cannot be overemphasized that geological maps and sections prepared from photographs and/or imagery alone may be grossly in error, and, planning of engineering projects involving geological considerations without careful field checking on the ground is at best a highly dubious procedure.

In recent years excellent results in reconnaissance geological investigations have come from the use of three systems of airborne instrumentation, in addition to conventional photography from the air. These systems utilize: (1) side-looking radar imagery (SLAR), (2) thermal radiometric imagery, and (3) color-infrared photography. Technical details of these systems are beyond the scope of the present treatment, and only brief summaries will be given.

Side-looking radar imagery (SLAR)

This technique employs a transmitter which generates short bursts or pulses of radio-frequency energy by means of a scanning directional antenna which confines the energy to a narrow beam.

Energy reradiated back to the antenna from terrain features passes through a cathode ray tube. Returning signals sweeping across the face of the tube are recorded on a continuous strip of film drawn across the face of the tube and provide an image of the terrain.

The images develop only from returned radiation, and, characteristically, in regions of considerable relief reveal "radar shadows" for terrane sloping away from the sensing device. To overcome this difficulty radar imagery requires multiple aerial traverses of the terrane in different directions.

The technique has a particular value in interpreting the configuration of ground and bedrock exposures in regions of very heavy forest cover, but does not lend itself readily to construction of scale-true maps, because of the difficulties of producing a controlled mosaic from images obtained by oblique sensing. However, if topographic maps produced by other methods or vertical aerial photographs are available, data from the radar images often can be transferred with good results.

Thermal imagery

Infrared radiation is long wave-length electromagnetic radiation outside of the visible range. All natural materials are capable of absorbing and/or radiating (emitting) infrared radiation. Transmission of infrared radiation through the atmosphere is limited by an absorption spectrum which, in the wave length range 0–25 microns (1 micron = 0.001 mm), typically shows nearly complete absorption of certain bands of wave lengths and a high degree of transmission of others. The widest band of transmission contains wave lengths between 8 and 13 microns, and it is this band which is most commonly used in remote thermal sensing of the earth's surface.

Thermal imagery obtained from airborne investigations utilizes a sensitive radiometric system which responds to very slight differences in incoming thermal radiation. An oscillating scanning system sweeps the earth's surface under the aircraft and translates differences in incoming radiation to a moving film. The result is an image which maps the lateral distribution of earth features which have different temperatures and thermal emissivities. A typical positive point of the film shows areas which are warmer than adjacent areas as white or light gray and colder areas in shades of gray or almost black. Standard procedure includes imagery just before dawn, and, later, imagery after the sun has had an opportunity differentially to heat materials of different temperatures and emissivities.

The method is particularly sensitive to differences in temperatures of soil or alluvium caused by differences in the temperatures of contained bodies of ground water, and is useful in identifying subsurface features such as fault zones saturated with ground water.

Color infrared photography

Ordinary color film is sensitive only to wave lengths (colors) in the visible range of the electromagnetic spectrum. Special color film with a high degree of sensitivity to wave lengths in the near infrared and used with selected filters produces highly unusual, contrasting color patterns that help to delineate by marked color differences geologic features that may not be apparent in conventional black and white or color aerial photographs.

CORE DRILLING

Introduction

Core drilling with diamond or tungsten carbide bits provides a direct and very useful means of determining subsurface geological conditions. Not only are cores of rock made available for field and laboratory examination, but a hole is provided for other kinds of investigations, including geophysical studies and water testing. Although large-diameter cores may be obtained for special purposes, most core drilling is done with bits that yield cores between 7/8 inch and 2–1/8 inches in diameter. Table XX summarizes standard core-bit diameters used in the United States and Canada, and indicates the diameters of the core and the hole obtained with each size of bit. Core recovery generally is much better with large core bits than with smaller bits.

Core drilling yields an essentially linear sample of bedrock and serves its most useful purpose in enabling three-dimensional interpretation of geological features

TABLE XX

Standard diamond core-bit diameters
(Diamond Core Drill Manufacturer's Association)

Size	*Core diameter* (inches)	*Hole diameter* (inches)
A. Standard bits for use with single tube core barrel in competent rock		
XRP	$^{15}/_{16}$	$1^1/_4$
EWX	$^7/_8$	$1^1/_2$
AWX	$1^1/_8$	$1^7/_8$
BWX	$1^5/_8$	$2^3/_8$
NWX	$2^1/_8$	3
B. Bits for use with double tube core barrels for drilling in soft, friable, or broken formations		
EWM	$^7/_8$	$1^1/_2$
AWM	$1^1/_8$	$1^7/_8$
BWM	$1^5/_8$	$2^3/_8$
NWM	$2^1/_8$	3

already identified at the earth's surface by geologic mapping and in exploring subsurface geology in areas of extensive subsurface cover. In the past core drilling on many projects was confined to exploration of tunnel portal approaches, and little or no drilling was done along or in the vicinity of the course of the tunnel. This practice is changing, partly because of the advent of mole tunneling, which requires particularly careful assessment of geological conditions, and partly because of a general trend toward increasingly detailed investigations of proposed tunnel locations prior to design and construction.

Core drilling is costly and careful judgment is required in determination of the locations and depths of test holes so that maximum useful geological and engineering information may be derived. Core drilling generally is justifiable during the feasibility and design stages of planning, although, in regions of extensive surface cover or complex geology, drilling during the reconnaissance stage of investigations may be indicated.

Core drilling beyond the heading of a tunnel under construction, especially in mole tunneling, is becoming standard procedure because unanticipated underground difficulties cannot be overcome as easily in mole operations as in conventional mining operations.

An indication of the data that can be obtained from exploratory core holes, especially those drilled from the surface, is given by the check list or procedures ordinarily followed during core drilling as summarized in Table XXI.

TABLE XXI

Check list of procedures to be followed during core-drill investigation

I. *Overburden*
Determine depth and composition. Note whether there is a water table. If so, measure and record its depth.

II. *Weathered zone in bedrock*
Note characteristics and depth in core, and, if rock is too soft to yield cores, systematically collect samples of cuttings or washings from drill water at collar of drill hole.

III. *Bedrock below weathered zone*
Carefully preserve all core that is recovered in a core box, being careful to maintain a proper sequence of cores as related to depth. Make particular notes of the following:

a. *Core recovery*
Determined by fitting broken pieces together to the extent possible and measuring length of core recovered compared with the drilling distance. Express core recovery as a percent or a ratio.

b. *Lithology of core samples*
Paying careful attention to mineralogy, fabric, and evidence of alteration. Select, as appropriate, samples for petrographic examination and determination of physical properties in the laboratory.

c. *Soft zones, yielding little or no core*
These are critical and indicate swelling or squeezing ground. Observe any tendency of the drill hole to squeeze during drilling. Collect samples of cuttings or washings from drill water for laboratory mineralogic analysis.

d. *Closely fractured zones yielding very poor core recovery*
Collect samples of washings at collar of hole for mineralogic analysis.

e. *Joints*
Examination of fragmented joint blocks or joints in core cylinders is critical. Note attitudes and spacings of joints with respect to long directions of core. Examine joint surfaces to determine whether they are rough or smooth and whether they have localized rock alteration.

f. *Small faults preserved in core cylinders*
Examine as for joints.

IV. *Ground water and water testing*

a. Note location and attempt to estimate amounts of ground water encountered in bedrock. Generally this is not possible unless ground water is under an artesian head.
b. Note location and estimate amounts of water lost during drilling.
c. Conduct tests of water losses at specified intervals with controlled pressures at collar of drill hole. These tests, together with observations of drilling water losses, are very important, and, in many instances, provide the only available basis for predicting ground-water conditions at tunnel level.

A valuable procedure for water testing utilizes a packer or packers which confine the water testing to definite intervals of the drill hole.

V. *Ground temperatures*
Some indication of above normal rock temperatures may be given by temperatures of returned drill water. If circumstances indicate the appropriateness, temperatures may be measured by a thermocouple lowered into the drill hole, filled with water, and allowed to remain until temperature readings stabilize.

TABLE XXI (continued)

VI. *Gases*
Returning drill water at the collar of the drill hole should be examined for bubbles of gases, especially methane or carbon dioxide.

VII. *Geophysical measurements*
These should be conducted when the core drilling is completed. Included, as appropriate, are electrical logging, resistivity logging, seismic downhole, or crosshole measurements. See text discussions.

VIII. *Special operations*
a. In most core drilling operations a crooked hole is bored. Generally it is advisable to survey the hole by any one of several available methods, so that accurate locations of critical geologic features may be made.
b. Oriented cores commonly are required to determine compass directions and dips of planar features observed in core cylinders. Oriented cores should be taken at sufficiently close intervals to provide controls for attitudes and directions of planar features such as foliation, bedding, jointing, and small faults. Unoriented cores from vertical holes enable measurement of dips but not the strikes of planar features. In unoriented inclined holes neither dip or strike can be identified in cores with any certainty.
c. Instruments have been devised to take photographs of the walls of core holes and to monitor drill holes with television. These instruments may be used for inspection of features of special interest, such as fractured zones, where core recovery is poor.

IX. *Drilling operations*
Commonly the driller maintains a log of the drilling operation. Whether he does or not, a careful record should be kept of the depth and nature of each drilling operation. Included should be notations of drilling times in rocks of different properties, locations of sections creating difficulty in drilling because of squeezing or closely jointed rock, locations of cemented and cased sections of the drill hole, and any other unusual aspects of the drilling operation.

X. *Assessment of efficiency of drilling operation*
There is little question that the amount of core recovery is a function of the care and efficiency with which the drilling operation is conducted. Perforce the geologist must make a qualitative judgment of the drilling operation to avoid erroneous interpretation of core recovery as related to subsurface conditions.

Core-hole logs

A complete core-hole log represents by graphical means and by explanatory descriptions all of the features of geological significance encountered in the drilling operation. A great many systems have been devised to plot logs and to include all pertinent necessary descriptive material. The particular system that is used is a matter of personal preference or is dictated by uniform procedures in an agency or company. Because of space limitations some judgment must be made as to what

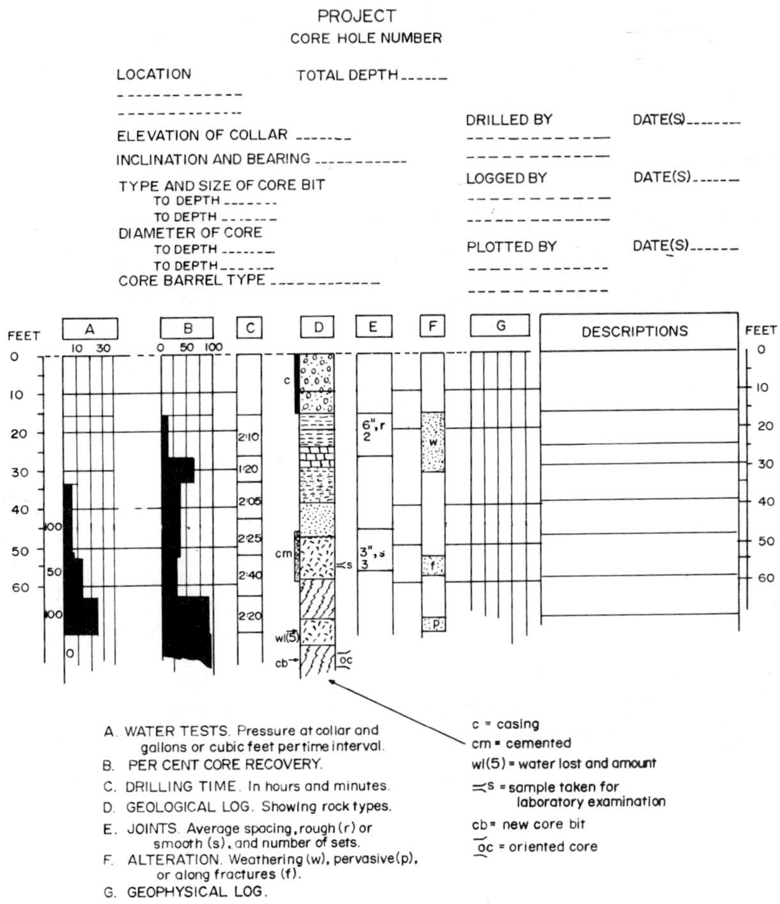

Fig. 81. Suggested form for plotting information from core drilling.

should be shown on the log and what should be included in the body of the accompanying report. A suggested form for plotting logs is shown in Fig. 81.

Water testing

Water testing should be considered as an integral part of the core drilling operation, especially as the drilling depth approaches the elevation of the proposed tunnel. A satisfactory method utilizes a perforated pipe and a packer which is set 10–20 ft. above the bottom of the drill hole. Generally, water tests with connections only at the collar of the drill hole are not particularly informative because of the difficulty of correlating water losses with the geological log.

Water tests should be performed for fixed intervals of time, say 5 or 10 min, with controlled pressures at the input at the collar, commonly of the order of 50–100 p.s.i. Losses are reported in gallons or cubic feet per specified time interval and plotted on the log.

Oriented cores

Many devices are available for extracting oriented cores during the drilling operation. Oriented cores are required by the geologist to measure true strikes and dips of inclined planar features such as foliation, bedding, and fractures. Commonly azimuths of known orientation are scribed on the core, so that upon removal from the core barrel the core can be rotated into its exact position relative to the rock from which it was extracted.

Strikes and dips of planar features are easily obtained by direct measurement from oriented cores from vertical or horizontal drill holes. However, strikes and dips from holes that have been drilled at an angle or have notably deviated from a vertical or horizontal direction during drilling require mathematical or graphical solution.

Solutions of true strike and dip of planar features in inclined drill holes is expedited by measurements of appropriate azimuths and inclinations of planar features and plotting them on a transparent overlay on a stereographic net. The equal-area net may also be used for solutions, but the writer prefers the stereographic net (Fig. 82) because elements of the net are circles, arcs of circles, or straight lines which facilitate construction.

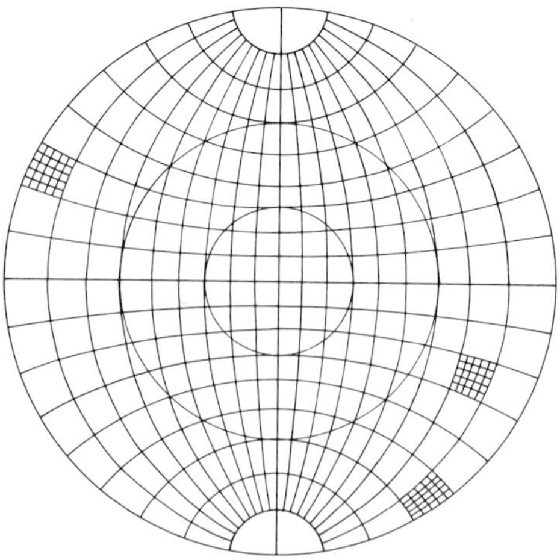

Fig. 82. Stereographic net obtained by stereographic projection on a diametral plane of lines of longitude and latitude from spherical projection.

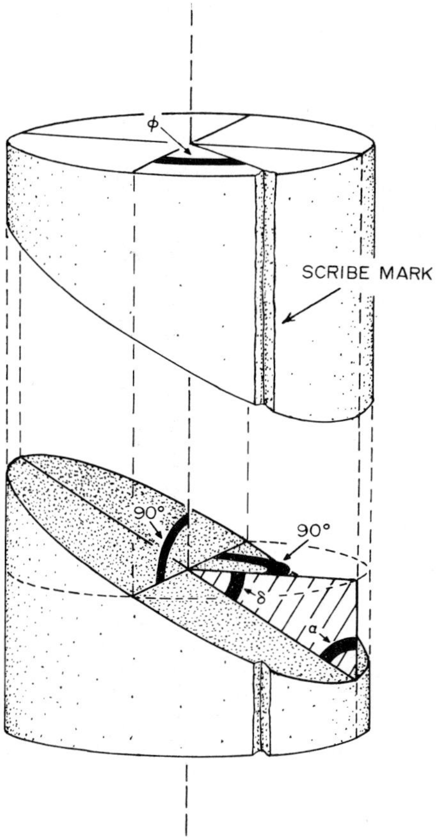

Fig. 83. Oriented core with axis in a vertical orientation. An inclined surface (heavily stippled) intersects the core and is inclined at an angle δ (or α) to the axis of the core. A horizontal line in the plane of the surface has an azimuth which makes an angle ϕ with the azimuth of a scribed line.

The use of the stereographic net for solution of strike and dip in a core from an angle drill hole is suggested in Fig. 83 and 84. Fig. 83 shows an oriented core intersected by an inclined plane (foliation, bedding, or a fracture). A scribed line provides an azimuth of known direction. The maximum angle that the inclined plane makes with the axis of the core is measured as α (alpha) or δ (delta) where $\alpha = 90° - \delta$. The azimuth, ϕ, of the inclined plane (a direction perpendicular to the core axis) is now measured relative to the azimuth of the scribe line. The azimuth is the direction of a horizontal line in the inclined plane when the core axis is vertical, and the maximum angle of inclination of the plane can be measured.

Fig. 84 illustrates the solution of an actual problem with a stereonet. A core hole drilled at an angle of 60° from the horizontal has a bearing of N59°E. An oriented

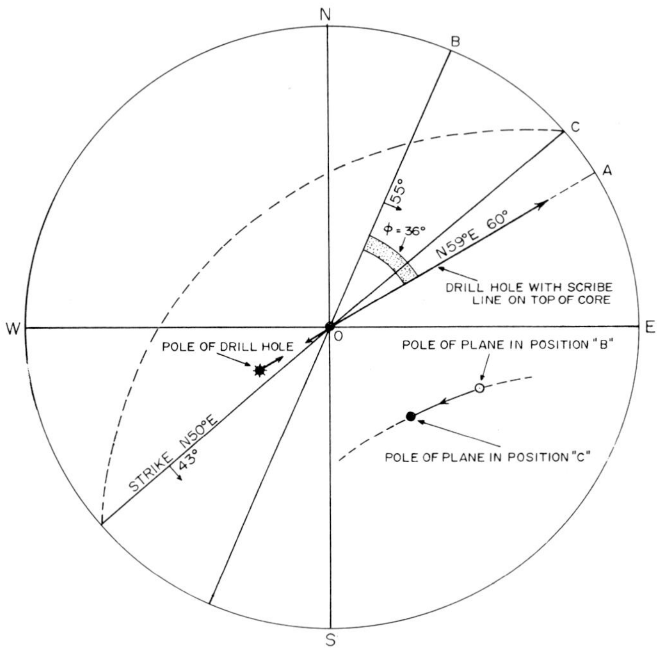

Fig. 84. Stereo plot (upper hemisphere) of an inclined drill hole.
Position *A*. Azimuth of scribe line on top of core when core axis is vertical.
Position *B*. Position of bedding plane on plot when core axis is not vertical.
Position *C*. Pole of core hole rotated to original position giving location of pole of bedding plane in its original orientation underground.

core with a scribe mark on the top proved to contain a bedding plane (it could have been foliation or a joint) with an azimuth $\phi = 36°$ with respect to the scribe line and an inclination of $\delta = 55°$ with respect to the axis of the core (see Fig. 83). The pole of the bedding plane (axis of core vertical) was plotted, and then rotated over the stereonet as the pole of the drill hole was rotated from the center of the plot to its original position. The pole in the new position yielded the desired solution, an actual strike of the bedding plane of N50°E and a dip of 43° to the southeast.

Geophysical investigation and instrumental logging of core holes

The petroleum industry and related service organizations have developed the theory and techniques of instrumental logging of drill holes to an advanced stage of usefulness. Many of these techniques have not been used routinely in geological investigations of engineering projects because in core drilling a continuous, coherent sample is available for examination, whereas in holes drilled for petroleum or natural gas only fragmented drill cuttings generally are available for inspection. Only two of the many available techniques have found wide utility in core-hole investigations, the resistivity and seismic techniques.

Resistivity logs of core holes are used in lateral correlation of layered materials of different resistivities and in the detection of permeable rocks, with either primary or secondary permeability, especially when the rocks are saturated with electrolyte-bearing water.

The use of seismic geophysical investigations of core holes is especially promising, and yields data that under favorable circumstances can be translated into engineering properties of rocks at depth. Core holes adapt well to determination of seismic velocities, either by recording down-hole seismic velocities by in-hole detectors and near-surface blasts, by continuous velocity logs, or by investigations utilizing two or more drill holes.

The relationship expressing the velocity of compressional seismic waves, in elastic materials, the waves used in seismic prospecting, is:

$$V_L^2 = \frac{E(1-\sigma)}{\rho(1-2\sigma)(1+\sigma)} \qquad (7\text{-}8)$$

where V_L is the velocity of the longitudinal (compressional) waves, ρ is density, E is Young's modulus, and σ is Poisson's ratio. It is clear that any natural process that causes a change in the elastic constants or density of a rock from its original condition will also cause a change in the velocity of propagation of the compressional waves.

Velocities in bedrocks depend on a large number of variables, including mineralogical composition, grain size, fabric, cementation, porosity, fluid content, anisotropy, temperature, confining pressure, and alteration. In practice and before interpretations of subsurface geology from seismic records can be made, every effort is directed toward determining average velocities of the various rock units within the area under study either by field or laboratory measurements. In the laboratory the most meaningful velocity measurements are made while the cores are subjected to confining pressures approximating those at depths at which the seismic impulses are transmitted by bedrocks in the field.

Correlation of seismic records with bedrock geology requires the closest cooperation between the geophysicist and the geologist. Seismic records which indicate velocities considerably less than usual or average velocities identified with each of the rock units suggest geological abnormalities that may be associated with wide fault zones, extensively jointed rocks, highly-porous materials, pervasive alteration, or the presence of previously unidentified low velocity undisturbed rocks between the geophones and the shot point. Obviously, all of these features are of concern in planning a tunneling operation.

Fig. 85 indicates some of the shot-point and receiver configurations that may be

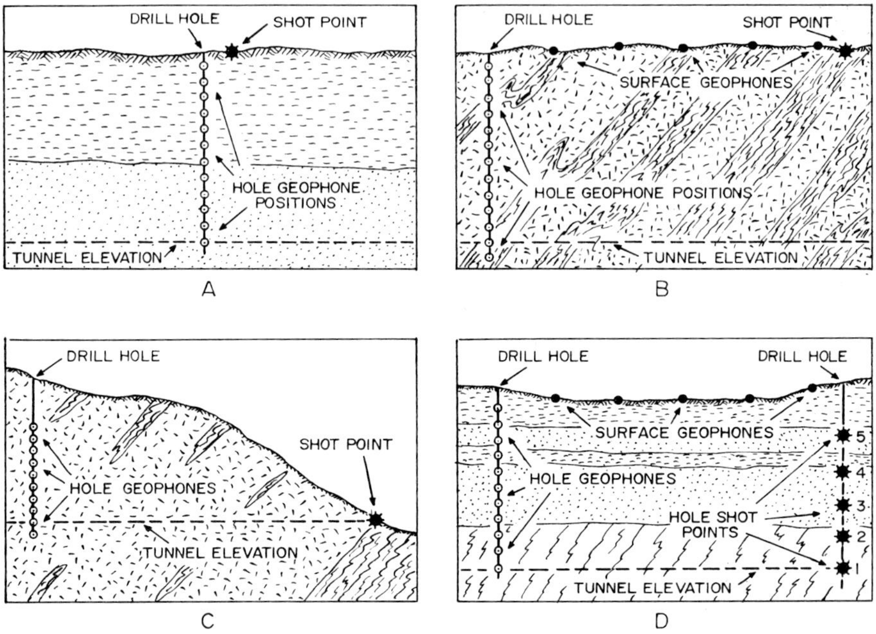

Fig. 85. Some possible configurations for seismic investigation of bedrock utilizing drill holes. A. Down hole measurement of velocities with a near shot point at surface. B. Bedrock investigation with hole and surface geophones and a distant shot point. C. Bedrock investigation in an area of considerable relief. D. Bedrock investigation using two core holes.

used in seismic investigations in bore holes. When such investigations are planned, it is advisable to drill holes with diameters provided by NWX or NWM core bits (hole diameter, 3 inches) so as to provide adequate working room.

USE OF EQUAL-AREA PLOTS FOR ANISOTROPIC ROCKS

The geological study of rock anisotropism in actual field examples is greatly expedited by use of statistical plots on an equal-area net (Schmidt net) in which the poles of planar features, such as bedding or foliation or faults or joints, are projected from the lower half of a spherical projection into the plane of the net.

Fig. 86 illustrates a construction for locating the pole of a plane of given strike and dip on spherical projection. The location of the pole can be characterized by an azimuthal angle (the strike) measured from the north direction, and by an angle between a north–south line and the normal to the plane (in this instance the dip angle).

The plane of equal-area projection is the horizontal equatorial plane of the

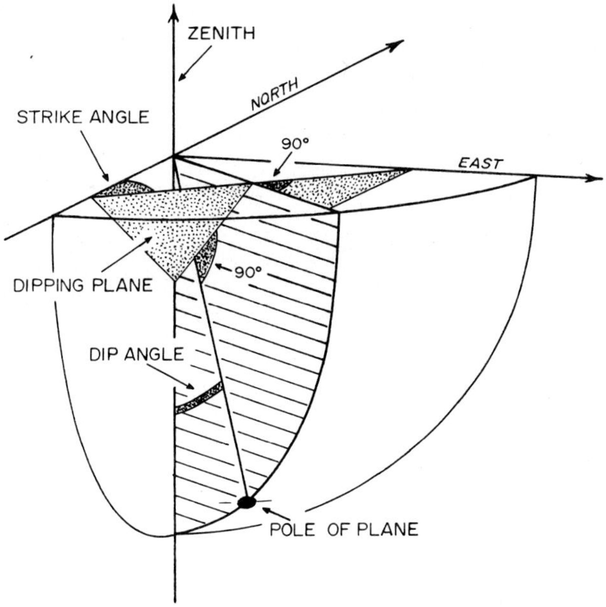

Fig. 86. Construction showing location of the pole of an inclined (dipping) plane on the lower half of a spherical projection. In the field the dip angle is measured from the horizontal in a plane of right angles to the strike of the plane.

spherical projection. Graphical projection into this plane of a pole on the spherical projection is indicated in Fig. 87. P is the pole of a plane on the spherical projection, and P' is its projection into the equal-area projection. The distance OP' from the center of the equal-area projection is related to the chord SP by the relationship:

$$OP' = \frac{SP}{\sqrt{2}} = \frac{SP}{1.414} \tag{7-9}$$

The equal-area net (Fig. 88) is constructed so that at a given latitude similarly enclosed areas are very nearly the same. On the north–south and east–west diameters (Fig. 88) the spacing of lines of latitude and longitude for a net with a 10-cm radius is given by the relationship:

$$r = 14.14 \sin \alpha/2 \tag{7-10}$$

where r is the radial distance from the center of the net, and α is an angle measured from the center of the net.

The use of the equal-area net is suggested in Fig. 89. The poles for planes, each

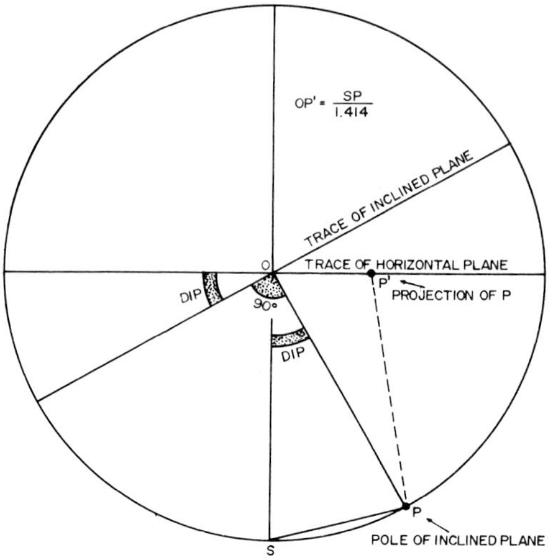

Fig. 87. Graphical projection of pole of inclined plane *P* into plane of equal-area projection *P'*.

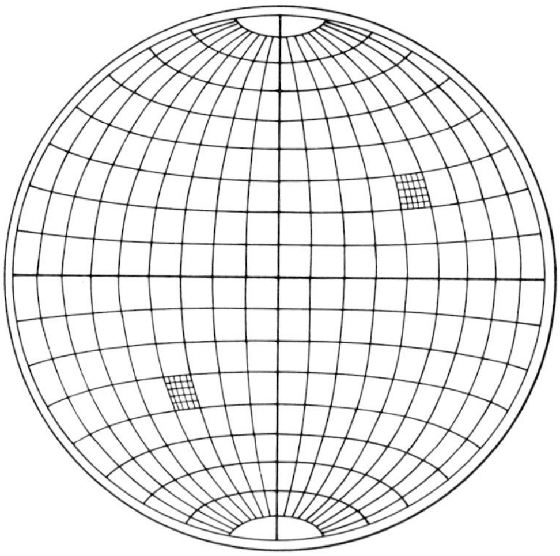

Fig. 88. Equal-area (Schmidt) net used in plotting poles of planes.

defined by a given strike and dip, are plotted on a sheet of vellum rotated over a net of 10-cm radius. With the aid of a 1-cm grid, a grid counter of 1 cm radius (area of counter equals 1% of total area within circle) the number of poles within the area of the counter is measured at each node of the grid and noted. With the aid of a peripheral counter numbers of poles at and near the circumference of the plot are determined. The numbers so obtained are converted into percentages of the total

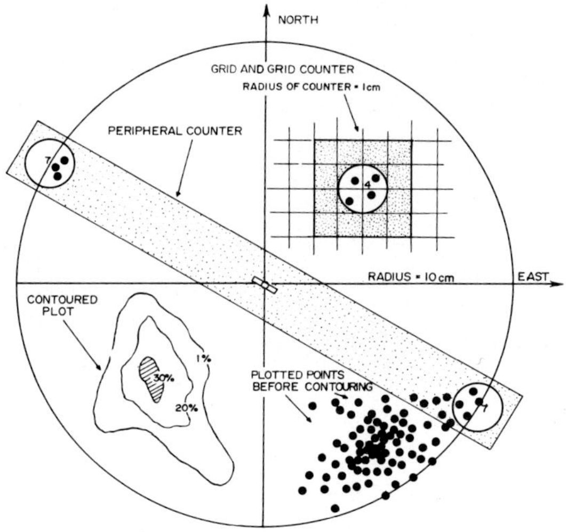

Fig. 89. Use of equal-area plot in statistical analysis of distribution of poles of plotted planes. Number of poles within area of a one per cent counter (radius 1 cm) is determined at nodes of a 1-cm grid with a grid counter and at the periphery with a peripheral counter. Per cent of total points is then determined for each number and contours are drawn through identical percentage concentrations.

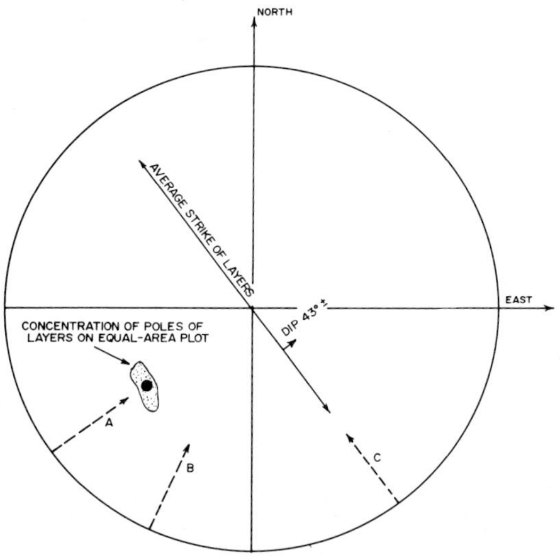

Fig. 90. The average strike and dip of a concentration of poles of layers on an equal-area plot as determined graphically with the equal-area net. *A*, *B*, and *C* are possible directions of tunnels intersecting the layered rock depicted in the plot. The appearances of excavated tunnel cross-sections corresponding to these directions are shown in Fig. 91.

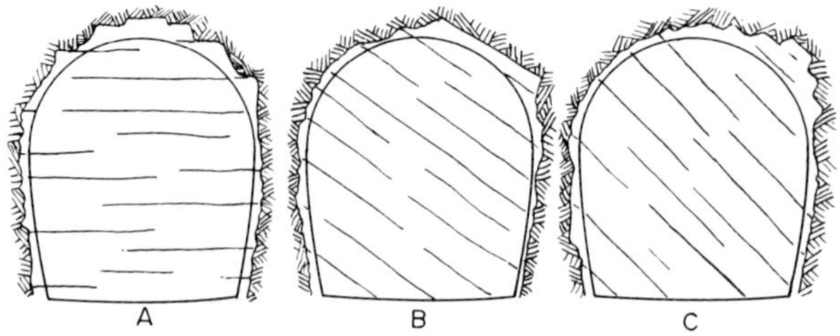

Fig. 91. Expected overbreak in tunnel cross-sections corresponding to directions of tunneling indicated in Fig. 84. The designed shape of the tunnel excavation is indicated to give some notion of the direction and magnitude of overbreak.

number of points and contoured. The 20-% contour in Fig. 89, for example, passes through all points on the grid for which the counter encloses 20% of the total number of points as the counter is shifted around the grid. Alternatively the counter may be moved around without reference to the grid to locate contours for various percentage concentrations more precisely.

A common application of an equal-area plot is indicated in Fig. 90. Three assumptions (*A*, *B*, and *C*) are made as to the direction that a tunnel might be driven with respect to the average attitude of layers in an anisotropic rock. The plot enables graphical solution of the apparent dip of the layering in the tunnel heading and suggests the probable typical configurations of tunnel sections (Fig. 91) excavated by drilling and blasting.

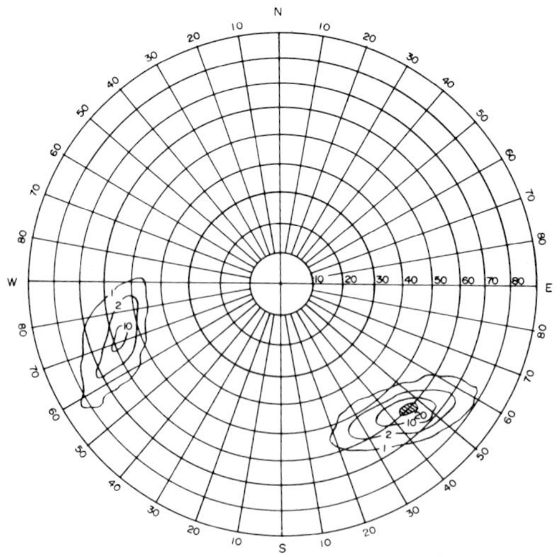

Fig. 92. Polar equal-area net from upper hemisphere of spherical projection. Contours show two concentrations of poles of a planar feature. The concentrations correspond to a strike of N45°E, dip 60°E, and a strike of S18°E, dip 48°SW.

Another method of plotting data on an equal-area plot is suggested in Fig. 92, which is a polar plot of concentrations of planar features. In contrast with the plot in Fig. 90 the projection is made from the upper hemisphere of the spherical projection, a strictly arbitrary procedure. This type of projection lends itself readily to computerization and is finding increasing use in the graphical representation of concentration of planar elements of rocks such as bedding, foliation, and joints. Instead of contouring by means of rectangular grid and a peripheral counter (Fig. 89) the count-out of poles of planes is made along radial lines at intersections of the radial lines with the circles of latitude.

STATISTICAL ANALYSIS AND COMPUTERIZATION OF GEOLOGICAL DATA

Large numbers of geological measurements of various kinds, as, for example, strikes and dips of joints or layering, are readily studied by the methods of statistical analysis and computer storage and retrieval. Many programs have been devised for computer analysis of geological data and for the rapid solution of many kinds of problems. The interested reader is urged to review the references listed at the end of the chapter.

REFERENCES

Dobrin, M.B., 1960. *Introduction to Geophysical Prospecting.* McGraw-Hill, New York, N.Y., 446 pp.

Grant, F.S. and West, G.F., 1965. *Interpretation Theory in Applied Geophysics.* McGraw-Hill, New York, N.Y., 583 pp.

Harbaugh, J.W. and Bonham-Carter, G., 1970. *Computer Simulation in Geology.* Wiley-Interscience, New York, N.Y., 575 pp.

Koch Jr., G.S. and Link, R.F., 1970. *Statistical Analysis of Geolocial Data.* Wiley, New York, 1: 375 pp., 2: 438 pp.

Krumbein, W.C. and Graybill, F.A., 1965. *An Introduction to Statistical Models in Geology.* McGraw-Hill, New York, N.Y., 475 pp.

Miller, R.L. and Kahn, J.S., 1962. *Statistical Analysis in the Geological Sciences.* Wiley, New York, N.Y., 483 pp.

Press, F., 1966. Seismic velocities. In: *Handbook of Physical Constants. Geol. Soc. Am., Mem.,* 97: 195–218.

Chapter 8

APPLICATION OF GEOLOGICAL INVESTIGATIONS TO TUNNEL PLANNING AND DESIGN

INTRODUCTION

Investigations of possible or proposed tunnel locations commonly are conducted at three levels of intensity:

(1) Reconnaissance investigations are made to provide a first general impression of the engineering and geological aspects of the proposed location and to establish a basis for planning subsequent investigations.

(2) Feasibility investigations involve more or less detailed consideration of site characteristics and preliminary consideration of design requirements and construction methods. Frequently, during feasibility investigations, the characteristics of alternate tunnel locations are compared, or a comparison is made of the economic practicability of the proposed tunnel as compared with other kinds of construction that might serve the same purpose. An end product of feasibility investigations of tunnel sites usually is a definite plan which may include a specifically identified tunnel site and appurtenant features.

(3) Investigations during the design stage are concentrated along a selected tunnel location and are specifically planned to provide the engineer, to the extent possible, with the kinds of information that he needs to design supports, auxiliary features, and to estimate quantities, rate of progress, and costs.

The cross-sectional size and configuration, and the amount and kind of tunnel lining are predetermined by the use to which the tunnel is to be put. Costs of labor, material, and equipment generally can be estimated with some precision, but, almost invariably, estimation of costs related to underground geological features must be made with a moderate to a great degree of uncertainty, especially in long tunnels in regions of geological complexity. Preliminary cost estimates made during the feasibility stage of study and final cost estimates made during the design stage in many large tunnel projects have proved to be grossly in error because of unanticipated or inadequately diagnosed geologic conditions.

Because of the great number of variables and because of the fact that in many situations subtle differences between rocks determine whether they behave competently or incompetently in a tunnel opening, computerization of geologic data ordinarily cannot provide an index of quality of rock that has any reliable significance. Accordingly, prediction of geological conditions at depth must to a large extent be based on intelligent guesswork based on experience and all available geological information.

In planning and designing a tunnel and in arriving at estimates of quantities and costs, it is essential that geological data be translated into some kind of quantitative or semiquantitative scheme of classification of competency or rock quality that can provide a basis for estimating variations in costs related to geological conditions.

CLASSIFICATION OF GEOLOGICAL MATERIALS

Throughout the ages wise men, individually and together, have expended vast amounts of time and effort in attempts to arrive at universally acceptable classifications of natural materials and processes. Changing theory, advancing technology, and individual biases have prevented the establishment of any widely accepted scheme of classification, and the geologist and the engineer are confronted with a bewildering variety of names, numbers, or letter designations for geological materials, fabrics, and processes.

The aim of geological classification is first to identify variables or parameters and then to devise a scheme of divisions and subdivisions that logically will place all similar materials or processes into definite, predictable compartments in the scheme of classification. With two variables, such as mineralogical composition and fabric, it is possible to devise a relatively simple petrographic classification that can be shown in a chart. Three variables can be represented by three-dimensional graphical models. However, in assessing competency of rocks in tunneling operations many variables must be taken into account, and any classification that is designed to preduce names, numbers, or letter designations that can be translated into cost estimates of tunneling must of necessity be highly vulnerable to error because much of the input into classification requires subjective evaluation of many of the variables.

Among the geological variables are time, origin, mineralogical composition, primary fabric, secondary fabric produced by earth dislocations, stored stresses, ground water, alteration, and depth of burial. To these must be added variables depending on engineering considerations, such as the size of the tunnel bore and the method of tunneling, before an assessment of competency can be made by any useable classification of earth materials.

A degree of assurance is derived from parameters or variables that can be measured in the laboratory or the field, and geological investigation of an engineering project should give maximum attention to such measurements, whether they are direct measurements of rock properties in the field or laboratory or numbers derived from geophysical exploration.

A widely used measure of rock quality (competency) is the *rock quality designation* (RQD) based on core recovery from exploratory drill holes (Deere, 1968; Deere et al., 1969). The aggregate length of pieces of hard and sound core 4 inches or longer is expressed as a percentage of the total length of run of the cored rock. RQD also may be estimated by geophysical measurements. The method has a particular application in estimating quality of rock in foundations for heavy structures and in short, high-cost tunnels of large diameter where concentrated core-hole investigation is economically justified. For investigation of long tunnels, particularly those at considerable depths, the method is less applicable because of the fact that the core sample obtained from one or several drill holes along the trend of the tunnel is so small compared to the total mass of rock being tested that it cannot be used as a base for extensive geological extrapolation, if only a RQD is calculated (see discussion of core drilling in Chapter 7). Cores obtained from drill holes are analogous to linear channel samples cut in ore bodies in mining operations, and the

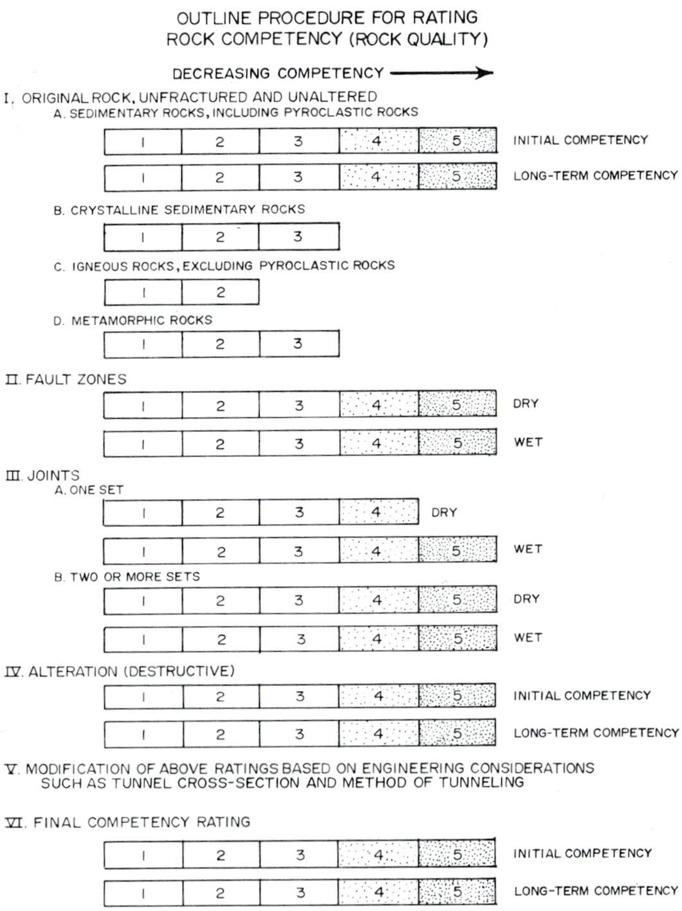

Fig. 93. Outline scheme for rating rock competency (rock quality).

TABLE XXII

Expanded outline procedure for rating competency (quality) of rocks at depth

---

Competency and incompetency are rated on a scale of 1 to 5 as follows:

1. Highly competent
2. Moderately competent
3. Marginal
4. Moderately incompetent
5. Highly incompetent

To arrive at a final rating of competency ratings are first made in individual categories as follows:

I. Original rock, unfractured and unaltered
II. Fault zones
III. Joints
IV. Alteration

Consideration is given in each of the categories to the effects of ground water on competency, and in category I to rock properties as they are related to depth of burial. As appropriate, ratings should take into account time-dependent competency, and separate ratings are indicated for short term initial competency and long-term competency.

A scheme for modifying ratings in categories I–IV, depending on engineering considerations, is given in category V, and in category VI a final competency rating is made after review of ratings and modifications in categories I–V.

I. *Original (fresh) rock, unfractured and unaltered*

Degree or rating of competency or incompetency of original rocks depends on origin, mineralogical composition, fabric, and depth of burial as an indicator of rock pressures. Rock pressures at moderate to considerable depths may cause dislocations in weak rocks, especially those prone to plastic behavior, that may not occur at shallow depths. Accordingly, the rating of competency should take depth of burial into account. Time is a factor in that failure of incompetent rocks in underground openings may be rapid, within a few hours or days, or delayed for a period of weeks to months. Allowances should be made in ratings for slow or delayed failure, or separate ratings should be made for rocks that fail quickly and those that are likely to fail after a prolonged period of time.

Ground water and moisture in the tunnel atmosphere contribute to the incompetent behavior of many materials, especially fractured materials, and ratings should include consideration of the effects of ground water on competency.

A. *Clastic sedimentary rocks including pyroclastic rocks*
Factors determining rating of competency are mineralogical composition, fabric, degree of lithification, amount and kind of matrix and/or cement, bedding, and fissility. Examples of incompetent rocks are mica-rich, poorly cemented siltstones, some shales and mudstones, and unconsolidated pyroclastic rocks.

Rate 1, 2, 3, 4, or 5 for initial competency and/or
Rate 1, 2, 3, 4, or 5 for long-term competency.

B. *Crystalline sedimentary rocks*
Nonporous crystalline limestone and dolostone and, possibly, rock salt.

Rate 1, 2, or 3.

TABLE XXII (continued)

Gypsum and anhydrite. Water may cause anhydrite slowly to hydrate and expand. A possibility exists that gypsum may dehydrate under strong unbalanced stresses.

Rate 3, 4, or 5 for long-term competency.

C. *Igneous rocks, excluding pyroclastic rocks*

Rate 1 or 2.

D. *Metamorphic rocks* such as slate, schist, and gneiss. Incompetency increases with depth in fine-grained to medium-grained schistose rocks containing abundant mica.

Rate 1, 2, or 3.

II. *Fault zones*

Tight faults with movement confined to a single surface rarely cause problems in underground openings, except when flat. Accordingly, the competency rating should be concerned especially with the materials in and associated with fracture zones of appreciable width localized along faults. Materials in fault zones are characterized by sheeting, breccia, or imbricate (braided) shears, and more or less gouge. Steep faults generally cause less trouble than flat faults, so that the attitude of a fault zone should be considered in the rating. Moisture is also a factor contributing to competency of the materials in and associated with fault zones, especially when gouge is abundant.

For dry fault zones:
Rate 1, 2, 3, 4, or 5
For wet fault zones:
Rate 1, 2, 3, 4, or 5

III. *Joints*

Joints exist in rocks in one or more sets, have various attitudes, and, if there is more than one set, intersect at various angles. Joint blocks display a variety of shapes depending on the angles of intersection and the number of joint sets. Important variables are attitude, spacing, and the texture of the joint surfaces whether rough or smooth.

A basis for rating incompetency in rocks without regard to surface texture or attitudes of joints based on average spacing follows:

1. Spacing > 10 ft. or > 3 m.
2. Spacing 3–10 ft. or 1–3 m.
3. Spacing 1–3 ft. or 0.3–1 m.
4. Spacing 2 inches to 1 ft. or 0.05–0.03 m.
5. Spacing < 2 inches or < 0.05 m.

Water in joints promotes incompetent behavior of jointed rocks especially when joint surfaces are smooth, and needs to be given consideration in establishing a rating.

When joints exist in a single set, attitude is an important factor in determining behavior of rocks underground. Steep joints generally cause less difficulty than flat joints, whether they are rough or smooth.

A. *One set of joints*
For dry jointed rocks:
Rate 1, 2, 3, or 4.
For wet, jointed rocks:
Rate 1, 2, 3, 4, or 5.

B. *Two or more sets of joints.*
For dry jointed rocks:
Rate 1, 2, 3, 4, or 5.
For wet jointed rocks:
Rate 1, 2, 3, 4, or 5.

TABLE XXII (continued)

IV. *Alteration (destructive)*

Alteration either reduces or increases the competency of rocks and commonly is associated with fracturing. Processes tending notably to decrease competency of rocks are argillization, chloritization, and sericitization, all of which usually cause a reduction in grain size and strength. Silicification, feldspathization, and sometimes dolomitization and calcification tend to improve the competency of rock. We shall concern ourselves here only with the processes that increase incompetency.

Alterations that reduce rock competency commonly produce materials that adsorb or absorb ground water to produce swelling rock, and incompetency in such rocks may not be immediately apparent because of a time lag after first exposure.

Availability of ground water or moist tunnel atmosphere should be taken into account when establishing a rating. Also consideration should be given to whether the alteration is incipient, moderate, abundant, or pervasive.

Rate 1, 2, 3, 4, or 5 for initial competency and/or
Rate 1, 2, 3, 4, or 5 for long-term competency.

V. *Modification of ratings based on engineering considerations*

A. *Modification of competency rating by giving weight to the cross-section of the tunnel*

The area and shape of the cross-section of the tunnel determine to a large extent the magnitude of excavation and support problems encountered in marginal or incompetent ground. There is no empirical relationship that can be used with confidence to establish a basis for modification of competency ratings with increase in the size of the tunnel bore.

If the rock encountered in the tunnel during full-face excavation is closely fractured and/or altered and collapses or is squeezed into the tunnel if unconfined by supports, it is evident that the forces causing collapse or squeezing are effective over a greater area in a large tunnel than in a small tunnel. Competency ratings of 3 or 4 for small tunnels probably should be changed to 4 or 5 for larger tunnels in the same materials.

To overcome severe problems of excavation and support in incompetent rocks, it is common practice to drive one or several smaller headings ahead of the full face to enable controlled excavation and placement of supports. Accordingly, any modification of competency ratings based on consideration of the size and shape of the tunnel cross-section should take into account the proposed methods for advancing the tunnel heading in difficult ground.

B. *Modification of competency rating based on method of tunneling*

With increasing use of tunneling machines (moles), consideration should be given to a comparison of the behavior of rocks excavated by the mole and by conventional methods (drilling and blasting).

Mole excavation produces a circular cross-section with maximum stability and does not shatter rocks to the same extent as blasting in conventional methods. Again there is no established empirical basis for modifying competency ratings based on a comparison of the relative effects on rock competency of the two methods of tunneling.

Because there is no blast damage in mole tunneling (although there may be sidewall damage in soft rocks from the pressure pads) a rating of 4 might be changed to 3 for a mole-bored tunnel of approximately the same cross-section.

TABLE XXII (continued)

VI. *Final competency rating based on review of ratings in categories I–IV*

Rate 1, 2, 3, 4, or 5 for initial competency and/or
Rate 1, 2, 3, 4, or 5 for long-term competency.

For engineering purposes the final rating may be evaluated as follows in terms of tunnel supports.

1. Highly competent rock requiring no support (Fig. 94).
2. Moderately competent rock requiring little or no support. Safety precautions may dictate need for support in short sections of tunnel.
3. Marginal rock requiring continuous light support for reasons of safety.
4. Moderately incompetent rock requiring continuous medium to heavy support and, for some tunnel sections, special tunneling operations such as grouting and breastboarding (Fig. 95–97).
5. Highly incompetent rock requiring very heavy support and special operations such as multiple-stage mining and extensive grout consolidation of rocks around tunnel and beyond tunnel face (Fig. 59).

In terms of rate of progress and the cost of the finished tunnel the rating may be interpreted as follows:

1. Maximum progress and minimum cost.
2. Progress slightly slower than in Rating 1 and costs slightly higher.
3. Rate of advance of tunnel heading slower than for Ratings 1 and 2. Costs higher because of requirement for installation of continuous light support.
4. Progress slow to very slow. Costs of excavation and support high.
5. Progress minimum and costs of excavation and support very high to extreme.

same statistical controls used in interpolation or extrapolation of data from channel samples apply to core samples.

From the above statements it must be abundantly clear that there is no classification of competency based on numerical parameters that can be considered as completely adequate under the circumstances that surround most tunnel site investigations, and frequently, the engineering classification of geological materials at depth must be based on subjective appraisal of a variety of variables and, in some instances, on geological intuition.

To assist in the sharpening of the appraisal of the several variables requiring subjective or objective analysis and to bring into focus the various factors that should be considered in estimating rock competency at depth, Fig. 93 and Table XXII were prepared. It will be obvious to the reader that explicit and detailed use of the competency rating schemes in Fig. 93 and Table XXII in many instances will require certain kinds of data that may not be available or obtainable. However, use of the rating scheme serves a special purpose in emphasizing the need for collection, to the extent possible, of data in several important categories during field and laboratory geological investigations of the locations of proposed tunnels.

The outline procedure in Fig. 93 should be used only after giving consideration to the various factors summarized in the several categories in Table XXII.

Fig. 94. Highly competent, unfractured biotite gneiss. Roberts Tunnel, Colorado. (Photo by Denver Board of Water Commissioners.

Final ratings of rock competency are made by reviewing competency ratings which depend on (*I*) the characteristics of original, unfractured rocks, (*II*) the effects of faulting in fault zones of appreciable width, (*III*) the influence of joints on competency, and (*IV*) the effects of rock alteration on the strength and behavior of rocks underground. In addition, consideration is given to engineering variables such as the size of the tunnel bore and the method of tunneling. A final rating is not simply an arithmetical average of the ratings in categories *I–IV*. Jointing and alteration in rocks with marginal ratings in each category may work together to produce incompetent rock. Accordingly, the final rating must be based on a careful estimation of relative contributions of each of the rated factors to the development of incompetency. Balanced judgment is required in making this estimation except when the rock initially is rated as moderately incompetent or highly incompetent in one or more categories.

Fig. 95. Incompetent interbedded shale and sandstone. Note breast-boarding and steel and timber crown bars installed above steel set as a safety measure before drilling. Pachecho Tunnel, California. (Photo by U.S. Bureau of Reclamation.)

Fig. 96. Incompetent swelling and squeezing ground in altered gneiss after 30 days of exposure. Note splintered timber lagging. West Heading, Upper Arch Heading No. 7. Straight Creek Tunnel, Colorado. (Photo by Colorado Division of Highways.)

Fig. 97. Incompetent, closely sheared shale. Note crown bars projecting beyond last steel set to protect workers. Roberts Tunnel, Colorado. (Photo by Denver Board of Water Commissioners.)

A special feature of the classification is provision for a final rating of initial competency, that is, competency at the time that rock is first exposed in the tunnel heading and long-term time-dependent competency, which may differ from initial competency. Examples of time-dependent competency are increase in strength by work-hardening in some plastic materials and decrease in strength of other rocks to which moisture has slow access to cause delayed swelling or slaking.

FEASIBILITY INVESTIGATIONS OF PROPOSED TUNNEL SITES

Under some circumstances the locations of the portals of proposed tunnels are precisely determined by geographic or other considerations and a study has established the fact that a tunnel is the only facility that will meet engineering needs. Examples are traffic tunnels or water tunnels in a heavily populated urban area. Additional factors may dictate only one possible alignment for the tunnel, and a possibility does not exist that the course of the tunnel between the portals can be altered to take advantage of favorable geological conditions. In this situation geologic investigations are immediately directed toward gathering of data that will be used in design, estimation of quantities and costs, and preparation of contract documents.

With less rigid restrictions on the location of the portals and the alignment of a tunnel, feasibility studies include comparison of cost-benefit ratios of alternate schemes of meeting engineering requirements. Feasibility studies based on geological factors assume a different hue. An example is the situation in which comparisons are made between the relative advantages of tunnels and cut-and-cover surface conduits for water transportation, and there is considerable flexibility in the location of tunnel portals and tunnel alignments. The problem for the geologist now becomes one of studying the geology of an area rather than the geology along a fixed tunnel alignment, so that he can assess the relative merits of alternative portal locations and tunnel alignments, examine surface features as they are related to possible construction of surface rather than underground facilities, and make appropriate recommendations to the planners.

Geological investigations at this stage are conducted by conventional methods employing aerial photography, remote sensing, available or prepared topographic maps, and careful ground checking and mapping. The result is a geologic map that shows the detailed geology of bedrock exposures, location and classification of surficial deposits, structural symbols identifying primary and secondary fabrics, and notations relating to ground-water characteristics and rock alteration. Geophysical investigations may prove to be valuable in interpretation of bedrock structure, and, in critical areas, core drilling may be necessary to arrive at an understanding of subsurface geological details.

Not uncommonly geologic maps of a particular area are already available. However, if these maps were not prepared specifically for use in the engineering appraisal of subsurface conditions they may not provide the kinds of information needed for adequate analysis of engineering properties of rocks at depth. Examples of maps commonly available are those prepared by Federal and State agencies and by students and faculty associated with educational institutions. Accordingly, there is, in most instances, considerable justification for a complete restudy of an area with the express purpose of special detailing of geologic features of engineering concern.

A geologic map of an area prepared during feasibility studies serves a very useful purpose in subsequent investigations when a tunnel alignment is finally established, particularly in identifying geologic features that project into the tunnel location but are not visible at the surface along the tunnel alignment because of heavy surface cover.

In making a selection of a particular alignment in preference to others several basic rules are followed:

(1) Tunnel portals should be located in competent bedrock or where excavation will not create instability in thick and/or extensive unconsolidated surficial de-

posits. If a tunnel approach in unstable materials cannot be avoided, provisions should be made for removal or for adequate containment of the unstable materials in the design of the tunnel.

(2) A tunnel alignment should be selected which intersects well-defined zones of incompetent rock associated with faults, closely-spaced joints, or layers of incompetent sedimentary rocks as nearly at right angles as possible. Avoid, if possible, alignments which pass through rocks containing flat faults or closely-spaced flat joints.

(3) Because canyons and valleys frequently are localized along fracture zones, now buried by alluvium and not visible at the surface, avoid tunnel alignments parallel or subparallel to and beneath the floors of canyons and valleys. An additional reason is that, if fractured rocks are present, heavy water flows, fed by streams or ground-water, may be encountered in the tunnel. If the tunnel must pass beneath canyons or valleys, try to select an alignment as nearly at right angles to their trend as possible.

(4) Jointed and/or altered rocks sometimes exist in zones with highly irregular and gradational boundaries. An example is provided by highly jointed and altered rocks within an igneous stock and in adjacent, surrounding country rocks. Such zones may be very extensive, and should be avoided during a tunneling operation to the extent possible.

Local topography and/or hydraulic requirements may dictate fixed locations for the tunnel portals. In this situation consideration should be given to the relative merits of a straight alignment between the portals as compared with a curved or dog-leg alignment which will avoid or minimize excavation difficulties in known or suspected intervals of incompetent or water-bearing rocks.

Determination that a curved or dog-leg alignment is preferable to a straight alignment must be made on the basis of careful cost analyses of alternatives. It is clear that lengthening of a tunnel by dog-legging can be justified only if the estimated total cost of the longer tunnel is demonstrably less than the estimated cost of a straight tunnel, taking into account the geologic differences of alternate routes.

Making estimates of costs of dog-leg versus straight alignments calls for very careful exercise of judgment regarding the geologic classification of underground materials as related to costs of excavation and support. Costs of tunnel driving in competent or marginal ground must be balanced against tunnel costs in incompetent ground, and the lengths in the alternate alignments of intervals of materials of various characteristics at tunnel level must be estimated as accurately as possible by projection, interpolation, and extrapolation of geologic and geophysical data from surface and core hole investigations.

In a typical comparison of alternate alignments, relative engineering costs can be estimated with reasonable accuracy if tunneling progress, the amounts and kinds of supports and the special operations required in boring through bad sections of rock can be assessed adequately. Estimation of costs necessary to overcome problems created by ground water is very difficult because there is no known method of accurately predicting water inflow. Apparently, the choice between alternate alignments must rest on overall as well as detailed consideration of the geology.

Because of many difficult problems in projection of surface geology to tunnel level, especially in deep tunnels, the geologist should be prepared to make an appraisal of the degree of certainty or uncertainty with which the projections are made. In situations where the choice between alternate alignments cannot be made with confidence, it may be necessary to undertake additional geological, geophysical, and core drilling operations in critical areas to obtain the required data.

Computer programs have been developed for calculation of tunneling costs based on geologic and engineering inputs. An example of computer analysis of the relative costs of alternate tunnel alignments has been described by Wheby (1971). Of course, computer evaluation of tunnel sites is no better or worse than the quality of the geological input, which, at times, must be highly subjective. The advantage of computerization is that it provides answers very quickly, whether they are good or bad, depending on the quality of the input.

DESIGN-STAGE INVESTIGATIONS

Design-stage investigations of proposed tunnel sites are special-purpose, additional investigations to assist the designing engineer in the preparation of detailed drawings and specifications and estimation of quantities and costs. Generally the specifications and estimates of quantities and all pertinent engineering and geological data, including those obtained during design-stage investigations, are included with contract documents for review by prospective bidders.

Geological investigations while final design of the tunnel is being completed are directed primarily toward improving the quality of geologic projections from the surface to tunnel level and to improve to the extent possible the engineering classification of various geologic materials that it is anticipated will be encountered as the tunnel-heading progresses. Investigations may include additional surface geologic mapping, geophysical work, and core drilling in critical areas.

The degree of confidence with which known geological features identified by surface geologic mapping and core-drill operations, with supporting data from geophysical studies, can be projected horizontally or vertically depends on a large number of factors. In engineering classification of materials and estimation of the presumed intervals in which each type of material exists along the elevation of the tunnel alignment, the geologist should indicate to the designing engineer the degree of confidence with which each projection is made. Of course, in regions of extensive surface cover, complex geology, and infrequent rock exposures, there may be

TABLE XXIII

Ratings of confidence in projection of surface geologic features laterally or to depth

I. *Projected with maximum confidence.*
   A. Single strong faults, fault zones, or jointed zones with sharply defined planar boundaries of known strike and dip as determined in surface exposures or drill holes.
   B. Unfolded and unfaulted sedimentary layers of constant thickness for which details of stratigraphy are available from surface or subsurface examinations.
   C. Layered pyroclastic or lava flow sequences in which thicknesses of individual layers do not vary appreciably from place to place.
   D. Monolithic, large igneous intrusions, especially large bodies of granite.

II. *Projected with considerable confidence.*
   A. Master faults of known attitude in regions of moderate geological complexity.
   B. Joint sets associated with master faults.
   C. Faulted or folded well-understood sequences of sedimentary rocks, pyroclastic rocks, or lava flows of constant thickness for which the geometry of faulting or folding is known.
   D. Thick, unfaulted or moderately faulted sequences of metamorphic rocks of known regional fabric.

III. *Projected with some reservations as to accuracy of projection.*
   A. Subsidiary faults in complexly faulted areas.
   B. Joint systems associated with subsidiary faults.
   C. Folded or faulted layered rocks for which fold or fault geometry is only partly known.
   D. Layered rock sequences containing angular unconformities.
   E. Highly lenticular layered sequences.
   F. Cross-cutting igneous intrusions such as dikes and stocks.
   G. Weathering localized along zones of fracturing.
   H. Destructive alteration at depth by warm to hot solutions of a variety of origins.
   I. Joint sets or systems not apparently related to master or subsidiary faults.
   J. Lenticular igneous intrusions in metamorphic rocks.
   K. Lenticular inclusions of metamorphic rocks in igneous bodies.
   L. Conformable igneous intrusions such as sills.

IV. *Questionable projections.*
   A. Faults or joint systems, major or subsidiary, not exposed at surface or encountered in drill holes and assumed to be present because of surface geophysical investigations or analysis of overburden or landforms.
   B. Altered zones of irregular shapes and unknown structural controls.
   C. Layered rocks with imperfectly known stratigraphy and attitudes.
   D. Surface features identified solely by float mapping of soil and rock debris.

V. *Highly questionable projections:*
   A. Projections based on hypothetical assumptions coming out of a knowledge of the geological history of an area.
   B. Projections based on theorizing by the geologist without adequate geological investigations in the field.

nothing to project except fragmentary bits of geologic information, various geological probabilities or possibilities, and information derived from exploratory drill holes and surface geophysical investigations. In such a situation close study of the distribution and configuration of the landforms and detailed examination of deposits of unconsolidated surface cover in a search for clues to bedrock structure and composition are imperative.

Table XXIII summarizes ratings of the confidence with which geological features observed at the surface or in core holes can be projected so as to construct longitudinal geological sections along the alignment of a proposed tunnel. The ratings assume that the features that are being projected have been identified in the field by direct observation or identified as probably or possibly existing, or by interpolation or extrapolation of bedrock exposures at the surface. Surface cover formed by local mass wasting in which rock fragments have not moved far from their bedrock source may be used in identifying lithologic variations in bedrock, and to a limited extent enables interpretation of features such as faults, joint systems, and alteration. However, bedrock features covered by overburden transported a considerable distance from the point of origin, such as in glacial and stream deposits, must be identified almost entirely by inference.

Accuracy of projection from the surface to depth decreases with increasing depth. The confidence ratings in Table XXIII assume moderate depths, say, of the order of a thousand to two thousand feet. Clearly, projection to shallow depths is likely to be more accurate than projection to greater depths.

Projection of geologic features into a vertical longitudinal section along a tunnel alignment is not simply projection downward or laterally of features observed along the alignment at the surface, but includes also projection into the section of features on either side of the alignment at the surface. Inclined planar features that are parallel or subparallel to the tunnel alignment ordinarily will not be intersected by the section at the surface and must be projected into the section from surface exposures on either side of the alignment. Accordingly, construction of the longitudinal section commonly requires knowledge of the three-dimensional aspects of the subsurface geology.

The object of surface geological mapping, geophysical studies, and core drilling is to gather as many data as possible in an effort accurately to classify materials and geologic conditions at depth. The magnitude of the effort varies considerably, depending on local geologic conditions. Of prime concern is the extent to which observed surface bedrock conditions provide an adequate sample of the kinds of geologic conditions that can be expected as the tunnel is excavated.

To provide a basis for elaboration of this point, several hypothetical longitudinal geologic sections are shown in Fig. 98. The sections are essentially bare-rock sections, and only minimum surface cover by overburden is assumed.

Fig. 98A illustrates a tunnel location in horizontal sedimentary rocks. Caprock

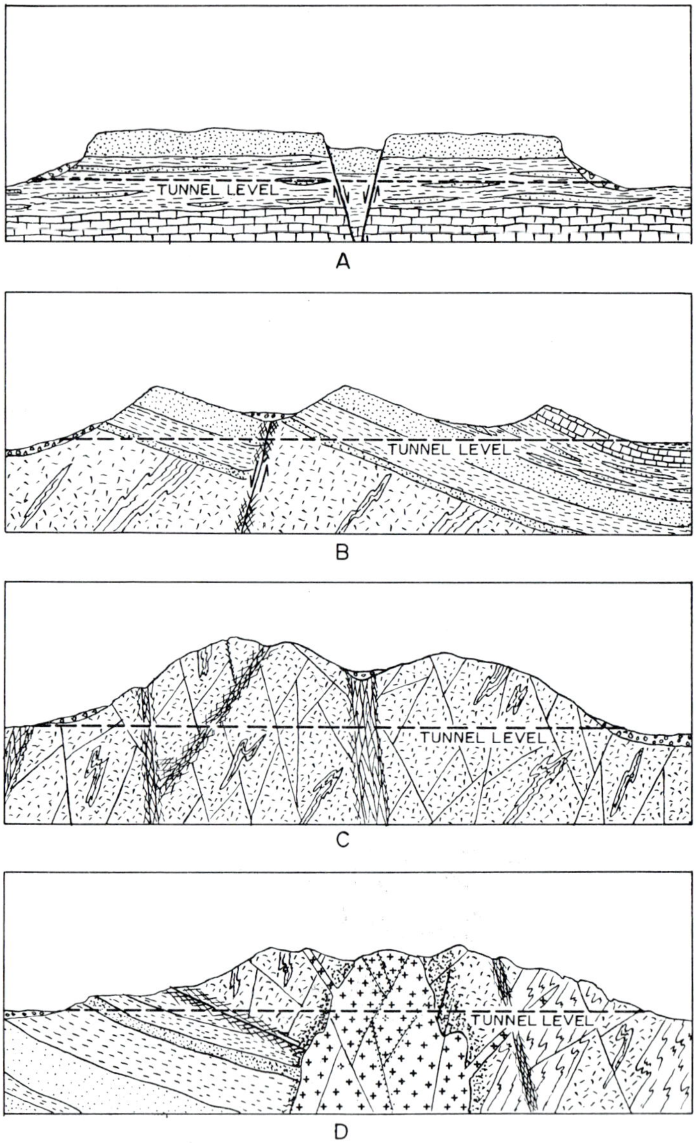

Fig. 98. Hypothetical longitudinal cross-sections along alignments of proposed tunnels. See text discussion.

consisting of sandstone is underlain by a succession of siltstones, shales, and lenticular sandstones, and, at greater depths, by limestone. Talus covers to a large extent bedrock exposures beneath cliffs formed in the caprock. Two faults intersect the sediments and project downward to tunnel elevation. Except for location and identification of the faults surface geologic mapping will reveal nothing concerning the rock conditions at tunnel level, except, possibly, in scattered exposures in talus beneath the cliffs at either end of the tunnel alignment. In this situation efforts are made to work out details of the stratigraphy below the caprock by regional studies, and by means of coreholes drilled through the caprock to tunnel level.

In Fig. 98B a succession of tilted sedimentary rocks rests unconformably on granite containing lenticular inclusions of schist and gneiss. A major fault accompanied by extensive jointing intersects and offsets the granite and overlying sedimentary rocks but cannot be inspected at the surface because of a blanket of overburden. Detailed surface mapping and measurement of a stratigraphic section, together with core drilling to provide additional stratigraphic details, and to determine the attitude and characteristics of the fault will enable accurate projection of surface features to tunnel level.

Fig. 98C shows a large mass of crystalline igneous rock containing lenses of schist and gneiss. Bedrock is intersected by numerous major and minor faults of a variety of attitudes. The major faults are accompanied by wide zones of fracturing, and, possibly, have localized considerable rock alteration. Detailed surface mapping, geophysical investigations, and core drilling in critical areas will reveal much concerning the general nature of conditions at tunnel level. Major fault zones, if strikes, dips, and widths are identified at the surface, can be projected to tunnel level with some confidence, but locations in the tunnel of smaller faults and larger faults not detected at the surface or of unknown strikes and dips cannot be determined. In this situation the mapping, geophysical studies, and core drilling are designed primarily to obtain a good sample of the *kinds of conditions* that exist at tunnel level, but, for most features, *not the exact locations.*

In calculating progress, quantities, and costs it is assumed that at tunnel level the proportionate total intervals of various kinds of rock conditions at depth will closely approximate the total intervals identified at the surface by surface investigations. It is certain that fractured zones will be intersected during the tunneling operation, but the precise locations, except for well-identified major features, will be highly uncertain.

Fig. 98D is a complicated geologic section along a tunnel alignment. Crystalline, igneous, and metamorphic rocks have been thrust over a succession of sedimentary rocks along a flatly-dipping thrust fault, and, later, an igneous stock with subsidiary dikes has intersected the rocks below and above the thrust fault. Faulting preceded and followed the intrusion of the stock, and extensive hydrothermal alteration is

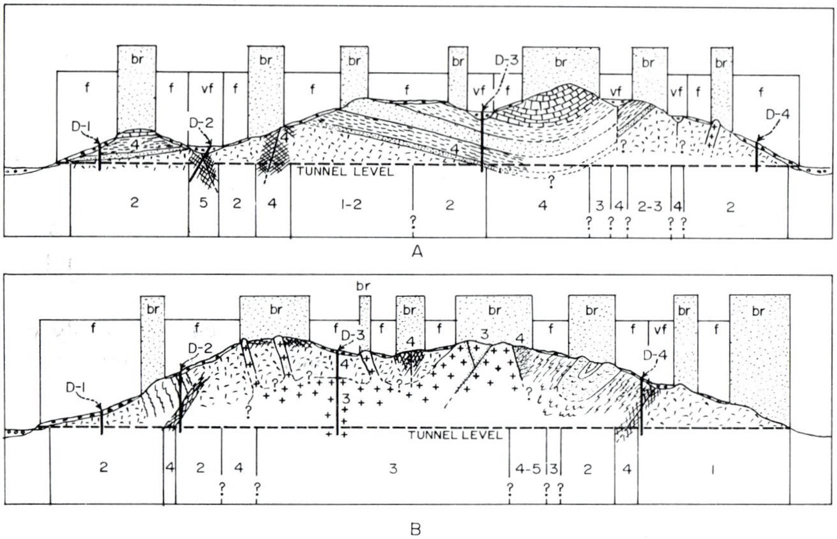

Fig. 99. Hypothetical longitudinal geological cross-sections of proposed tunnels with locations of bedrock exposures (*br*), float cover (*f*), and valley fill (*vf*) at surface. Exploratory core holes are indicated as *D-1,D-2,* etc. Competency of rocks at tunnel level, based on projections, is rated on a scale of 1–5 (see Fig. 93 and Table XXII).

irregularly localized around the stock (stippling) and along fractures within the stock.

The thrust fault, major steep faults, and the layered sedimentary rocks can be projected to tunnel level with some confidence, but many other features, such as the downward extension of the stock, the attendant alteration, and the minor faults, cannot be projected satisfactorily.

Design stage estimates can be in part based on safe projections, but, to a large extent, must reflect uncertainties as to geological conditions along major intervals of the tunnel. Nevertheless, careful surface investigations provide the *only* basis for design-stage estimates, and should be implemented to a maximum degree so as to provide the best geological sample of detailed and general conditions that is possible under the circumstances.

Fig. 99 illustrates hypothetical examples of geological sections used as a final basis for estimating competency of rocks at tunnel level. The sections clearly identify all features, surface and underground, that enabled projections to tunnel level of various geological features, including the locations of bedrock (*br*) exposures at the surface and core-drill data. A distinction is made between surface cover that permits identification of the lithology of the bedrock (float; symbol, *f*) and valley fill (*vf*), which reveals nothing concerning underlying geologic features. Numbers below the tunnel level and elsewhere are competency ratings according to the scheme outlined in Fig. 93 and Table XXII, and provide a means for estimating tunnel costs for various intervals. Question marks are used to indicate uncertainties in geological projection.

In Fig. 99A a series of folded and faulted sedimentary rocks rest on a basement of granite. Surface exposures and exploratory drill holes permit reasonable projection to tunnel level. A layer of poorly-cemented shales and siltstones above a basal sandstone is judged to be incompetent and is a matter for concern, as well as well-identified faults accompanied by extensive fracturing. According to usual practice, faults or suspected faults of unknown dips are shown as being vertical, leading to some uncertainty as to the locations of the faults at tunnel level.

The situation illustrated in Fig. 99B requires considerable guesswork as to locations of certain subsurface elements. An igneous stock has intruded faulted crystalline igneous and metamorphic rocks. Extensive fracturing and alteration characterize the contact zone in the country rock adjacent to the stock and rocks in this zone are rated as incompetent. The configuration of the stock at depth is unknown, but dikes exposed at the surface give a clue as to its probable location.

Although the drawings in Fig. 99 do not provide an analysis of probable groundwater conditions, rock temperatures, possible noxious gases, or other aspects of concern in the tunneling operation, data summarizing these features normally are included on the drawings or in accompanying descriptive text materials.

A CASE HISTORY

A measure of the uncertainties of projection of surface geological features to tunnel level in a mountainous area of complex geology is provided in a study of the

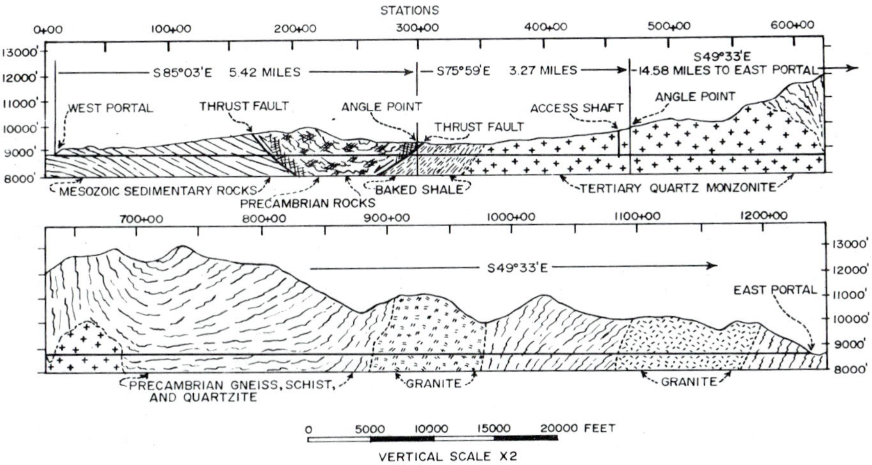

Fig. 100. Generalized section through the Harold D. Roberts Tunnel, Colorado.

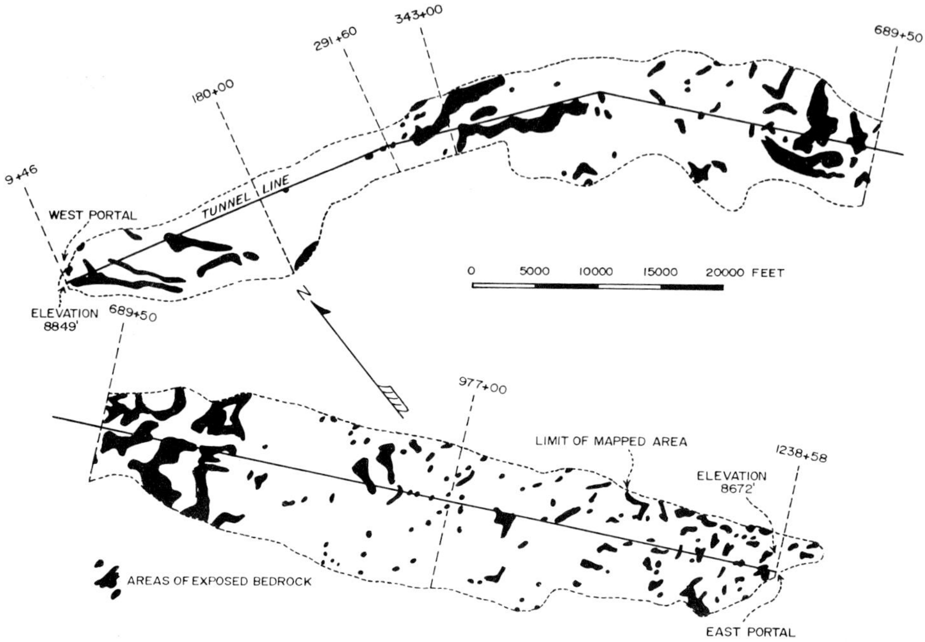

Fig. 101. Map showing areas of exposed bedrock. Harold D. Roberts Tunnel, Colorado.

23-mile-long Harold D. Roberts tunnel in the Front Range of Colorado (Wahlstrom, 1964). Feasibility-stage studies included consideration of several alternate alignments utilizing geologic maps along the alternate alignments, geophysical investigations and logs from a total of 6,357 ft. of exploratory core drilling. Additional geological mapping at the surface after the completion of the tunnel provided an unusual opportunity for comparison of surface conditions with those actually encountered in the tunnel (Wahlstrom and Hornback, 1962; Warner and Robinson, 1967).

Fig. 100 shows a highly-generalized longitudinal geological section of the tunnel, and Fig. 101 indicates the areas of exposed bedrock at the surface in a strip map along the tunnel alignment. Surface cover consists of products of mass wasting on the slopes and mountain tops, and glacial and stream deposits in the valleys, and within the map area conceals 87% of the bedrock. Float mapping in several covered areas enabled approximate delineation of boundaries of several lithologic units but revealed little concerning joints and minor faults in bedrock.

Projections of rock bodies of simple geometry and planar structures such as large faults from the surface to tunnel level prior to construction of the tunnel proved to be fairly accurate. However, extensive surface cover prevented an analysis of detailed fabrics. A knowledge of the general geology of the area in the vicinity of the tunnel provided a basis for supposing that joints and faults would be abundant in

TABLE XXIV

Numerical data, Harold D. Roberts tunnel

	Station 9+46 to 180+00	*Station 180+00 to 291+60*	*Station 291+60 to 343+00*	*Station 343+00 to 689+50*	*Station 689+50 to 977+00*	*Station 977+00 to 1238+58*	*Summary 9+46 to 1238+58*
Maximum depth of tunnel below surface (ft.)	1,090	1,020	690	4,290	4,450	2,360	4,450
Minimum depth of tunnel below surface (ft.)	0	550	550	540	1,250	0	0
Area mapped at surface (sq. miles)	3.84	2.12	0.82	10.54	10.69	8.68	36.69
Per cent of mapped area containing areas of bedrock exposures	10.9	3.3	16.7	17.6	14.4	8.6	13.0
Estimated per cent of actual bedrock exposures	6.7	1.9	8.6	6.4	5.7	2.4	5.1
Number of faults and veins mapped at surface	13	4	13	85	30	11	156
Strong faults and veins mapped at tunnel level	62	41	15	111	32	39	300
All faults and veins mapped at tunnel level	688	1,037	189	1,420	376	196	3,906
Average spacing of all faults and veins at tunnel level (ft.)	24.8	10.7	27.2	24.4	76.5	133.5	31.47
Strong joints mapped at tunnel level	1,292	0	312	2,372	908	421	5,305
Average spacing of strong joints at tunnel level (ft.)	13.2	–	16.5	14.6	31.7	62.1	21.1*
Number of mapped faults and veins intersecting tunnel line at surface	12	0	2	13	15	8	50
Ratio of mapped faults and veins directly above tunnel to strong faults and veins at tunnel level	12/62=0.19	0/41=0	2/15=0.13	13/111=0.12	15/32=0.47	8/39=0.20	50/300=0.17
Ratio of mapped faults and veins directly above tunnel to all faults and veins at tunnel level	12/688 =0.017	0/1037 =0	2/189 =0.01	13/1420 =0.09	15/376 =0.04	8/196 =0.04	50/3906 =0.012
Estimated per cent of bedrock actually exposed along tunnel line at surface	6.9	2.3	21.1	7.4	8.8	1.4	6.5

* Excludes interval between stations 180+00 and 291+60.

certain sections of the tunnel, but no surface controls were available for even approximate projection of joints and faults from the surface downward for most intervals of the tunnel.

Table XXIV provides a numerical comparison of faults mapped at the surface and faults encountered in the tunnel. The numbers of strong joints mapped in the tunnel also are indicated, except for a segment of the tunnel from Station 180+00 to Station 291+60 above a thrust fault. Joints in this segment were so closely spaced and complex that a count could not be made.

REFERENCES

Deere, D.U., 1963. Technical description of rock cores for engineering purposes. *Rock Mech. Eng. Geol.*, 1: 18–22.

Deere, D.U., 1968. Geological considerations. In: K.G. Stagg and O.C. Zienkiewicz (Editors), *Rock Mechanics in Engineering Practice.* Wiley, New York, N.Y., pp. 1–20.

Deere, D.U., Merritt, A.H. and Coon, R.F., 1969. *Engineering Classification of In-situ Rock.* Rep. for Air Force Spec. Weapon's Center, Kirtland Air Force Base, New Mexico.

Wahlstrom, E.E., 1964. The validity of geologic projection, a case history. *Econ. Geol.*, 59: 465–474.

Wahlstrom, E.E. and Hornback, V.Q., 1962. Geology of the Harold D. Roberts Tunnel, West Portal to Station 468 + 49. *Geol. Soc. Am. Bull.*, 73:1477–1498.

Warner., L.A. and Robinson, C.S., 1967. Geology of the Harold D. Roberts Tunnel, Station 468–49 to East Portal. *Geol. Soc. Am. Bull.*, 78:87–119.

Wheby, F.T., 1971. Computerized cost estimates as an aid to tunnel route planning. *Bull. Assoc. Eng. Geologists,* 8: 165–181.

Chapter 9

TUNNELING METHODS

INTRODUCTION

Until recently tunnels in rock were driven exclusively by so-called conventional or classical methods, that is, by drilling and blasting. As required, supports were installed to contain bad sections of ground and to provide safety for the workers. Special operations included sealing-off of water flows by grouting and installation of tunnel linings appropriate to the designed function of the tunnel.

In the 1950's James S. Robbins and Associates developed a successful hard-rock tunneling machine that was used in Canada, and a new era of mechanization of tunneling was born. Since then, numerous companies in the United States and abroad have developed tunneling machines, popularly called "moles", and many examples of high-speed, economical tunneling with moles have been cited in a rapidly growing literature.

Geological and engineering studies of proposed tunnels now routinely include comparison of relative costs of mole tunneling versus tunneling by conventional methods, and, in some instances, bidders on construction of a particular tunnel are given an option as to the method of tunnel excavation.

The geologist, as well as the engineer, is confronted with an entirely new set of problems in the feasibility and design stages of investigation. In tunnels driven by conventional methods a working face with adequate room for dealing with adverse underground problems is always available. A mole at a tunnel heading is a heavy, massive device practically filling the tunnel for the full length of the machine, and rapid advance into swelling, squeezing, or highly-fractured ground, especially water-bearing ground, can result in total and disastrous immobilization.

Clearly, the geologist who is concerned with the progress and safety of a proposed mole tunneling operation may find it necessary to increase his perception of the geology of the proposed tunnel site greatly if he is to do more than estimate the proportions and kinds of geologic conditions that will be encountered. Now, he finds that, to the extent possible, he must determine *where* the conditions exist and make a critical judgment as to *how bad* they are. Information much more detailed than might be required for a tunnel to be driven by conventional methods may be necessary to make a choice between mole and conventional methods of excavation.

A glossary of terms commonly employed in tunneling operations is presented at the end of this chapter.

CONVENTIONAL TUNNELING METHODS

Conventional tunneling methods, also called "classical" methods, are cyclic operations involving, in sequence, drilling, blasting, mucking, and installation of required supports. The cycle is interrupted for special operations such as grouting, or spiling through very difficult ground.

In large tunnels much of the activity at the tunnel heading is centered around a jumbo, a platform or a tier of platforms mounted on wheels or skids. The jumbo is used during drilling operations, loading of drill holes with explosives, and provides working access to the arch and walls of the tunnel as supports are installed.

In most modern operations tunnels in competent rock generally are driven "full-face", even when the tunnels have large cross-sections, as in highway tunnels. However, especially in the late nineteenth and early twentieth centuries, large tunnels were driven by a series of small tunnel excavations, called "drifts", in various portions of the larger tunnel heading.

Multiple-stage advance of tunnel headings is today used in very difficult or incompetent rock. It is well known that caving, running, or heavy ground is much more easily managed and contained in small openings than in large openings, and excavation of one or several small heavily-supported tunnels sometimes provides the

Fig. 102 Top drift in squeezing ground in closely sheared pulverized gneiss. Harold D. Roberts' Tunnel, Colorado. (Photo by Denver Board of Water Commissioners.)

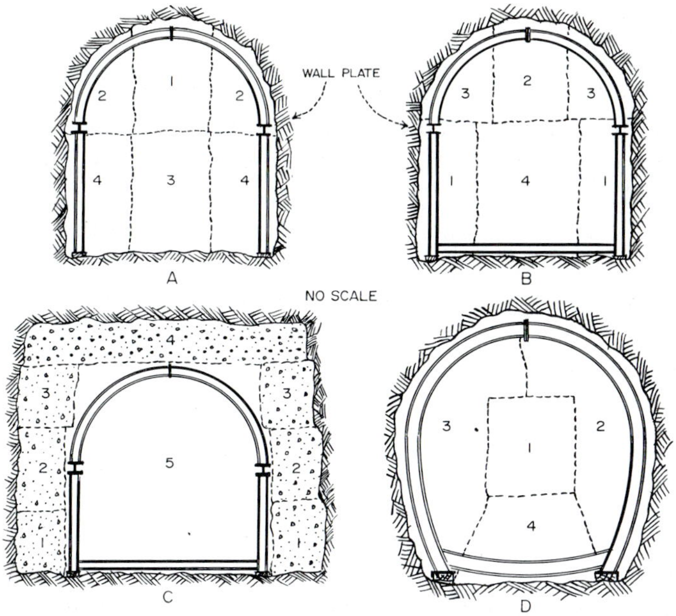

Fig. 103. Several methods of advancing a tunnel heading by drifts in incompetent rock. Sequence of excavation is indicated by numbers. Temporary supports installed in drifts are not shown. A. Excavation sequence started with a top drift. B. Excavation sequence initiated with side drifts. C. Drifts are driven outside of limits of tunnel and filled with concrete. D. Center drift is expanded to enable placement of steel ribs.

only feasible means of penetrating sections of highly-incompetent rocks (Fig. 102). In some situations stabilization before advancing the tunnel heading requires extensive grouting, or drifts which are filled with concrete around and outside of the planned tunnel opening.

A few of many possible methods for penetrating highly-incompetent rock are indicated in Fig. 103. In Fig. 103A a top drift is excavated (*1*) and with appropriate installation of temporary supports is expanded laterally (*2*), to enable placement of arch ribs or wall plates. This operation is followed by driving a bottom drift (*3*) which is widened (*4*) to install wall ribs below the arch ribs.

In Fig. 103B the wall ribs are placed within side drifts (*1*) and capped with wall plates. Leaving a central core (*4*) excavations (*2,3*) are made to install the arch ribs. Finally, the core is removed, and an invert strut is bolted or welded to the wall ribs.

In extremely incompetent rock it may be required to drive a succession of drifts and fill them with concrete, as in Fig. 103C, before excavating the full cross-section of a tunnel. A capping beam (*4*) is constructed of reinforced concrete to support the rock load in the tunnel arch.

Fig. 103D shows a method for installing steel supports in a tunnel of intermediate size. A heavily supported central drift (*1*) is expanded by excavation and installation of additional supports until the ribs can be placed.

TUNNELING MACHINES

Tunneling machines have been used to excavate tunnels with diameters of about 6 ft. to more than 36 ft. Rates of excavation of over 400 ft. per day have been recorded in soft ground. Excavation at the rate of about 100 ft. per day in moderately hard limestones and sandstones is routine. Rate of progress, other factors being equal, is proportional to the strength, toughness, and hardness of the rocks intersected in the tunnel, and it is common practice to correlate progress with the compressive strengths of rocks. Advances in technology now make mole excavation possible in rocks that have compressive strengths of 20,000 p.s.i. and greater, although most successful operations in the past have been in rocks with compressive strengths of 15,000 p.s.i. or less.

Tunneling machines generally are custom-built for a particular project, and the design includes consideration of the kinds of geologic conditions that are anticipated. Important structural elements of a typical mole include a rotating cutter head, and provisions for controlling forward thrust and alignment. A canopy or

Fig. 104. Hughes Tool Company tunnel boring machine used in driving Tunnel No. 1 on the Navajo Indian Irrigation Project, New Mexico. (Photo by U.S. Bureau of Reclamation.)

shield system provides protection from falling rocks and enables safe installation of tunnel supports near the face of the tunnel. Cuttings from the face are transported to the rear of the machine by a variety of devices. Forward advance is controlled by a laser beam which provides precise alignment and grade for steering the machine.

Under ideal conditions mole tunneling in competent rocks is a continuous, non-cyclic process interrupted only for replacing cutters and for making mechanical repairs and adjustments. Advances in design are leading toward machines which permit adequate working room for installation of supports and easy access to the working face ahead of the tunneling machine when adverse conditions are encountered.

Fig. 104 shows a tunneling machine and Fig. 105 is a photograph of the tunnel excavated by the machine. The machine in Fig. 104 is only one of a great variety of tunneling machines, each with particularly advantageous design characteristics. Some machines are specifically intended for excavation in soft rocks such as shale. Others are constructed for excavation in very hard rocks such as granite, quartzite, and similar rocks with compressive strengths in excess of 20,000 p.s.i.

Fig. 105. Tunnel No.1, Navajo Indian Irrigation Project, New Mexico, looking upgrade toward tunnel boring machine. Note 4-inch I-beam roof supports with protective lagging. (Photo by U.S. Bureau of Reclamation.)

TUNNEL SUPPORTS

The history of the development of various techniques for installing supports in tunnels, especially in incompetent rocks, is a fascinating tribute to the ingenuity of man in overcoming seemingly impossible obstacles (Sandström, 1963; Szechy, 1967). A vast literature reviews and summarizes the theory and practice of safe containment of rocks with varying degrees of competency and under small to great load pressures in underground openings (see especially Proctor and White, 1968).

Through the ages timber and masonry supports in various configurations served their purpose well, but today steel, concrete, and mortar are the materials most used in construction of tunnels with uses projected over long periods of time. Timber is still used extensively in mining operations, where permanent support is not necessary, and in tunneling operations constructed with steel, concrete, and mortar as temporary supports. Timber has an unquestioned use because of the ease with which timber members are dimensioned and shaped to meet special purposes.

Because of the limited scope of the treatment in this book, no attempt is made

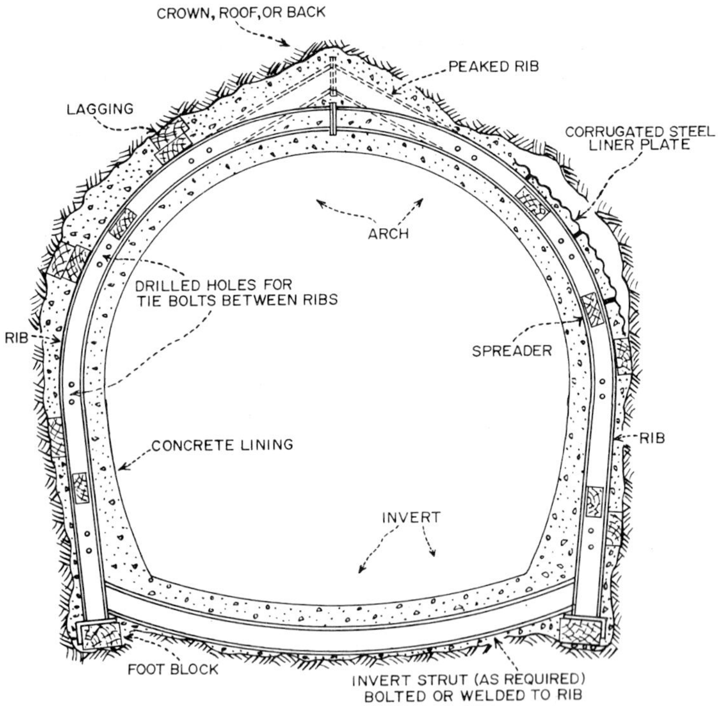

Fig. 106. Identification of support and lining components of a tunnel excavated by conventional mining methods. Peaked ribs provide space for pumping concrete into forms for formed linings.

Fig. 107. Steel mats and rock bolts in arch of a mole-bored tunnel in shale and sandstone. Sandstone layer is at springline elevation with shale above and below it. Note slaking of shale below the sandstone layer. Starvation Tunnel, Bonneville Unit, Central Utah Project. (Photo by U.S. Bureau of Reclamation.)

to outline the specifications for the great variety of materials and structural members used in supports in modern tunneling operations. Instead, only a brief summary is given so that the reader will have a general notion of various factors that are given consideration in the installation of a safe tunnel lining.

Materials commonly used in construction and lining of tunnels (and other kinds of underground openings) excavated in rock fall into several categories, as follows:

(1) *Steel ribs,* also called *bents,* which give rigid to semirigid support. The ribs are made from I-beam or H-beam structural steel bent to conform to the requirements of a particular tunnel cross-section. Fig. 106 shows a type of rib commonly used for support in a tunnel of horse-shoe-shaped cross-section excavated by conventional methods. Ribs may be "peaked" to make space for a pipe (called the "slickline") used to supply pumped concrete to forms used to place tunnel lining. An invert strut is installed when side or bottom pressures are active and provides a continuous ring of steel around the tunnel opening.

Fig. 108. Intake tunnel of pumping Plant No.1, Southern Nevada Water Project. Note use of shotcrete in marginal blocky ground. (Photo by U.S. Bureau of Reclamation.)

(2) *Timber members* are used in a variety of ways (Fig. 106) to provide propping or temporary support, and to provide a protective covering above the ribs. Where possible, timber members are removed before placing permanent concrete lining.

(3) *Steel liner plates.* These are placed outside of the ribs to prevent movement of rock materials into the tunnel opening (Fig. 106). In some tunnels steel liners are placed inside an already formed concrete lining so as to give additional strength. Commonly a liner placed in this manner subsequently is coated with mortar to prevent corrosion.

(4) *Rock bolts.* These are bolts with expansible anchors which are set in drilled holes in the arch or walls of a tunnel to improve rock competency. Sometimes the rockbolt holes are filled with grout during or after placement of the bolts.

(5) *Wire mesh or perforated steel mats.* Heavy steel wire mesh and steel mats are used to contain fragmented rock in the tunnel arch (Fig. 107). They are pinned in place with short, grouted steel pins or with rock bolts.

(6) *Miscellaneous steel members.* Steel members of a variety of shapes and dimensions are used for special purposes. Included are tie bolts, steel lagging, and

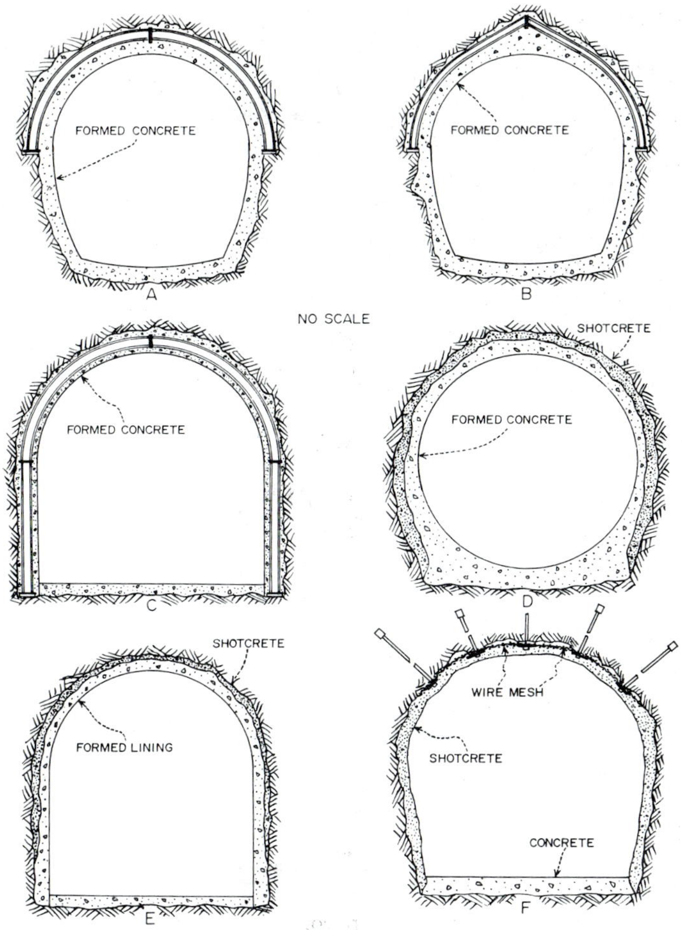

Fig. 109. Several possible cross-sectional configurations of lined and supported tunnels driven by conventional mining methods. A. Horse-shoe section with arch rib supports and formed concrete lining. B. Horse-shoe section with peaked arch rib supports and formed concrete lining. C. Stragiht leg and circular arch ribs with formed concrete lining. D. Pressure tunnel with initial shotcrete wall and arch support. Circular cross-section is completed by formed concrete. E. Shotcrete and formed concrete lining in a haulage or highway tunnel. F. Rock bolts, wire mesh (or steel mat), and shotcrete in a tunnel finished with a concrete floor.

units such as rails, rods, or channel-iron used in spiling through caving ground.

(7) *Gunite* is a form of plaster (mortar) consisting of fine-sand, cement and water which is shot at the arch or walls of a tunnel by air pressure to provide waterproofing and, in some instances, structural support. Commonly rock-bolted or pinned wire mesh is installed before applying the gunite.

(8) *Shotcrete* is a quick-setting thick plaster or mortar shot at rock exposures. A

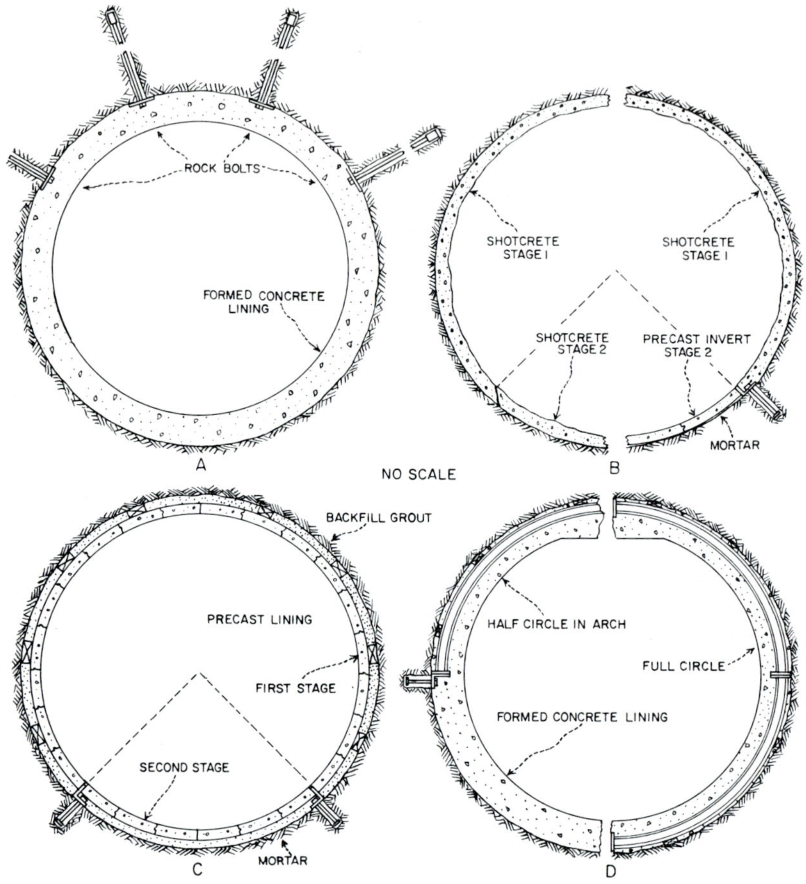

Fig. 110. Some possible cross-sectional lining and support configurations in circular tunnels driven by a tunneling machine. A. Rock bolts and formed concrete lining in a pressure tunnel. B. Shotcrete around entire periphery and shotcrete and preformed blocks in a finished tunnel. C. Preformed concrete blocks installed with backfill grout and mortar. D. Arch rib or full-circle rib and formed concrete lining.

mixture of cement and aggregate up to three-quarters of an inch are applied pneumatically and are mixed with a rapid hardening agent at the nozzle of a spraying device. The shotcrete leaves the nozzle at a velocity of about 500 ft./sec and establishes a firm bond with the rock, even though it is wet. In practice the shotcrete is placed in two inch layers until a desired thickness is attained.

Shotcrete provides an early flexible support which adjusts to minor dislocations of tunnel arch and walls and, in many situations, prevents formation of an arch which is initiated by spalling or initial detachment of joint blocks or slabs. Because of a rapid increase in strength over a short interval of time and ease of application shotcrete has particular use in mole-driven tunnels either to replace more conventional supports such as steel ribs and lagging or, in some instances, to provide a lining for a finished bore. An example of the use of shotcrete is shown in Fig. 108.

(9) *Grout* is a mixture of cement and water forced into rocks around the tunnel periphery to add strength to fractured rocks and intercept water flows.

Several possible cross-sectional configurations of lined and supported tunnels are indicated in Fig. 109 and 110 and suggest the many choices that are routinely available to tunnel designers.

GROUND WATER IN TUNNELS

There is no known method for estimating exact locations and volumes of underground water flows that might be encountered in rock tunnels. In regions of simple geology where the rocks are impermeable and unfractured and a steady surface supply of water is absent, the prognostication of a dry tunnel is relatively safe. However, as the complexity of the geology, as expressed in folding, jointing, and faulting increases, the uncertainty as to probable ground-water conditions increases. Ground water in limestone poses an especially uncomfortable prospect because of the possible existence of highly-irregular solution cavities and caverns partly to completely filled with free-flowing water (see Chapter 6).

Heavy inflow of water into tunnels almost invariably is associated with fissures in brittle, competent rocks and in many tunnels is encountered in the least expected places. Geological projections assist in locating zones of fractured rocks that might be water-bearing, but in a typical situation the exact flow-paths followed by underground water within a fractured zone are very intricate and completely unpredictable.

In addition to the necessity for removal of water by pumping, underground water contributes to the incompetent behavior of a variety of materials and presents a whole spectrum of problems during lining of a tunnel, whether by shotcrete or by formed concrete. Wise engineering practice dictates that every reasonable attempt should be made to seal off water flows by grouting before they are encountered in the tunnel excavation.

Depending on the area of cross-section of the tunnel and the degree of expectancy that groundwater might be encountered, one to several feeler holes should be drilled in continuous overlapping sequence beyond the tunnel heading. If water is encountered, it should be sealed off with grout. Large flows should be treated with special care, and additional grout holes should be drilled as required. Feeler holes are useful not only for detecting potential underground water flows but also enable assessment of rock conditions beyond the tunnel heading. Grouting of water-bearing zones generally increases the competency of the rocks in the zones as well as reducing or eliminating the inflow of water into the tunnel when the zones are intersected.

In conventional mining methods feeler holes are drilled up to 50 ft. in depth with a pneumatic drill at the same time or prior to the drilling of blast holes from the drill jumbo. In mole mining a single central hole is easily drilled ahead of the tunnel face while the mole is in operation, probably with a butt (noncoring) diamond bit. Drilling of peripheral holes requires a shut-down of the mole operation, but in water-bearing rocks the delay in excavation usually is abundantly justified.

Except in very-small-diameter tunnels a single feeler hole is not sufficient. The attitudes and locations of water-bearing fissures beyond the tunnel heading generally are unknown, and a dry single hole proves very little if it does not intersect a water course. In large-bore tunnels multiple, overlapping feeler holes should be drilled, if there is any reason to suspect the presence of water beyond the tunnel face.

A system of drilling feeler holes for a tunnel of about 10 ft. in diameter is suggested in Fig. 111. For larger openings more holes might be necessary. In Fig. 111 drill-hole patterns are alternated, and each pattern is drilled deep enough to extend well beyond the heading from which the next pattern will be drilled. When water is encountered in a drill hole a nipple and valve is placed in the collar of the hole, so that pressure and volumes may be measured and grout applied as required under controlled conditions.

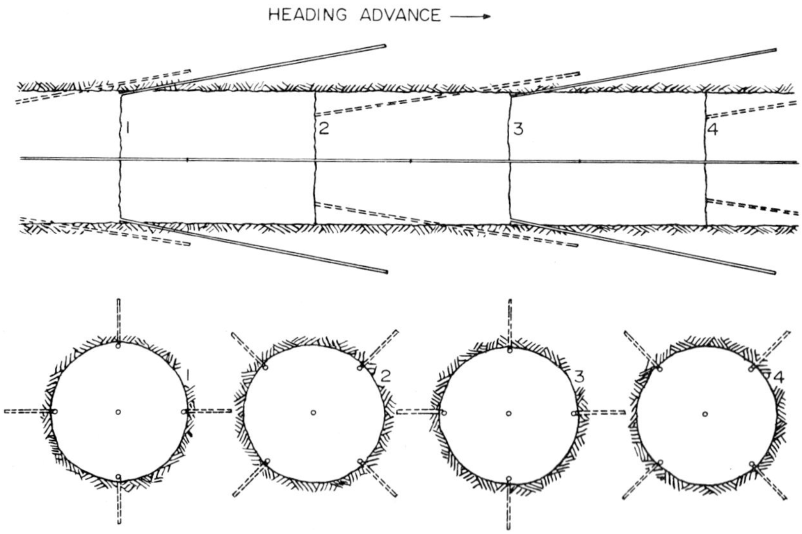

Fig. 111. Patterns for feeler-hole drilling beyond a tunnel heading. Numbers identify locations of tunnel headings when each pattern is drilled. Top view is either a vertical or horizontal section along center line of tunnel. Bottom views show successive locations of feeler holes in tunnel headings.

Fig. 112. Bulkhead of concrete-filled sacks installed prior to grouting to seal off heavy water flow estimated at 1,300 g/m. Flow was encountered in fractured, brittle, quartzitic sandstone. Roberts Tunnel, Colorado. (Photo by Denver Board of Water Commissioners.)

Grouting is an art. Many kinds of materials have been injected into grout holes to seal off water and to increase the competency of fractured rocks. Injected materials include various organic and inorganic chemicals (chemical grouting) and quick-setting plastics and cements. However, in general practice grout usually is prepared by mixing cement with water. Thin mixtures of about one sack of cement (1 cu. ft.) per 7–10 cu. ft. of water are initially introduced into the grout holes. Gradually the mix is thickened and pumping is continued to refusal. Grouting pressures must exceed the pressure of the water in the hole, but should not exceed a safe limit which is determined by the strength and competency of the rock. Probably pressures in grouting should not exceed 1000 p.s.i. under any circumstances, because grout under high pressure in fissures is entirely capable of producing new cracks where they did not previously exist.

Grout mixtures follow channels of easiest flow, and ordinarily will not penetrate clay or gouge-filled fractures. Accordingly, grouting is not particularly effective in sealing off slow, seeping water flows or increasing competence in gougy, thoroughly altered, or initially soft, clayey rocks.

Once ground water has been intersected by a tunnel, perhaps as a complete surprise, an entire set of new problems is created. The severity of the problems depends on a variety of factors including the volume of the water and whether it is associated with swelling or running ground, or is confined to one or a few fissures in competent rock.

Large or small flows may be supplied by relatively confined bodies of stored or trapped water, and the inflow of water may diminish rapidly with time. In some circumstances, especially when clayey or gougy materials are traversed by ground water the rate of flow gradually decreases because of self-grouting, whereby fissures eventually are plugged by sand-size particles and clay. In any event, heavy flows of water should be observed for an interval of time before remedial measures are taken to seal them off by grouting.

Grout sealing of heavy water flows in a tunnel at times is very difficult and not always successful. In typical procedures several holes are drilled to intersect the water course and grout, sometimes containing sand or sawdust, is pumped into the grout holes. To confine the spreading action of the grout, cracks from which water is issuing in the tunnel may be closed with oakum, lead wool, or wood wedges.

Under extreme conditions it may be necessary to build a bulkhead completely filling the tunnel behind the water course before proceeding with the grouting operation. An example is shown in Fig. 112.

A tunnel lining, either formed concrete or rockcrete, placed in tunnels where water is entering the tunnel opening, requires installation of drainage permitting flow of the water into the tunnel. Commonly so-called "weeps" are installed so as to permit the free flow of ground water that might be trapped behind the tunnel lining and exert disintegrating pressures.

CONVENTIONAL VERSUS MOLE TUNNELING

The impetus in modern tunneling is toward increasing use of tunneling machines. However, before a tunneling machine is specified for a particular operation, a careful comparison should be made between the advantages of driving the tunnel with a machine as compared with tunnel excavation by conventional methods, that is, by drilling and blasting. In some situations consideration should be given to employing both methods, taking advantage of especially applicable techniques of each method. Any decision that is made with regard to the method of tunneling ordinarily is based on engineering and geological factors which determine the costs of tunneling. However, special circumstances may dictate the use of a tunneling machine, as, for example, in a densely populated or industrialized area where blasting in a shallow tunnel might cause extensive damage to surface structures and underground utilities.

Advantages to be gained by using a tunneling machine, assuming that geological conditions are not adverse, are as follows:

(1) A smooth circular cross-section is obtained. Such a section has maximum stability and creates the most favorable distribution of stresses in rocks around the tunnel. Less support is required than in drilled and blasted tunnels.

(2) There is no blast damage and rock around the tunnel is left essentially intact.

(3) Overbreak associated with conventional tunneling is absent or minimal, and the volumes of concrete or rockcrete used in linings is much less than in a typical drilled and blasted tunnel.

(4) In suitable rock the rate of progress of the tunneling operations is much greater than in conventional tunneling operations.

(5) Blast damage to buildings or underground facilities in industrial and residential areas is eliminated.

(6) Less manpower is required in a mole operation than in cyclic, conventional tunneling.

Factors limiting the use of tunneling machines include the following:

(1) The initial cost of a tunneling machine is high, much more than equipment required for conventional tunneling. Accordingly, tunneling machines generally are not practicable for short tunnels or tunnels of very large diameter and intermediate length unless such tunnels are located in areas of population or industrial concentrations where conventional mining might lead to damage suits.

(2) Large and complex ventilation systems may be required to reduce dust and heat and to provide suitable working conditions near the tunnel heading.

(3) Control of alignment and grade may be difficult in soft sections of rock.

(4) Working room at the tunnel heading is very limited as compared with the available room in conventional tunneling methods.

(5) Rapid adjustments of techniques and equipment to changing rock conditions cannot be made with the same ease as in conventional methods.

(6) Tunneling machines generally are designed for the specific requirements of a particular project. In the absence of an accurate knowledge of geological conditions it may not be possible to design and construct a high-performance machine that will be versatile enough to meet unexpected demands of rapidly changing underground geological conditions.

(7) Costs of tunneling with moles in very hard rocks are high and may not be competitive with costs of conventional tunneling.

(8) The mobility of tunneling machines is much less than equipment used in conventional tunneling. This mobility is an important factor when a short turning radius is required or access to a tunneling operation is by shaft or a subsidiary bore.

(9) Tunneling machines yield poor performances where very hard rocks alternate with soft rocks or where extensive zones of highly altered and fractured rocks containing swelling, squeezing or running ground are encountered.

Analysis of the above statements leads to one outstanding conclusion. In proposed tunneling operations, in the planning and design stages as consideration is given to the relative advantages and disadvantages of mole excavation versus conventional excavation, it is essential that the decision as to which method is employed should be made *only after completion* of a thorough, professional investigation of the geological conditions along the tunnel alignment.

ACCESS TO TUNNELING OPERATIONS

Under some conditions, especially in regions of extensive surface cover and complex geology, preliminary investigations may include driving of a small-diameter exploratory tunnel. Such a tunnel is subsequently enlarged, or it may be used as access to a larger parallel tunnel driven from multiple headings. In very long tunnels, shafts may be sunk at one or more intermediate locations along the tunnel alignment to enable excavation of the tunnel at multiple headings.

TUNNELING TERMINOLOGY

A special language has become identified with tunneling operations. A summary of the meanings of frequently used words or expressions is shown in the accompanying Glossary.

Glossary of tunneling terms

Arch. Curved roof of an underground opening. The portion of the tunnel above the center line.

Arching tendency. Tendency of incompetent or moderately competent rocks to form an arch above an underground opening by caving. Commonly the arch assumes the shape of a Gothic arch in cross-section.

Back. Also called the "crown" or "roof". The highest portion of a tunnel.

Bar (bar down). Remove loose or insecurely attached rock fragments or slabs from arch of tunnel immediately after blasting and before placing supports.

Blocking. Wood blocks placed between steel supports and walls or arch of tunnel to provide support, and maintain alignment of ribs.

Blocky rock. Jointed rock that breaks into large or small blocks.

Bottom heading. A pilot-bore driven ahead of and at the bottom of a larger excavation.

Brattice. A wall constructed in a tunnel to deflect air toward the work face of a tunnel construction.

Breastboards. Wood planking placed temporarily at tunnel heading to contain incompetent rock (Fig. 113).

Bulkhead. A barrier placed at a tunnel heading to prevent inflow of water or shattered rock into the tunnel.

California switch. A movable double-track assemblage that enables switching of muck cars near the tunnel heading.

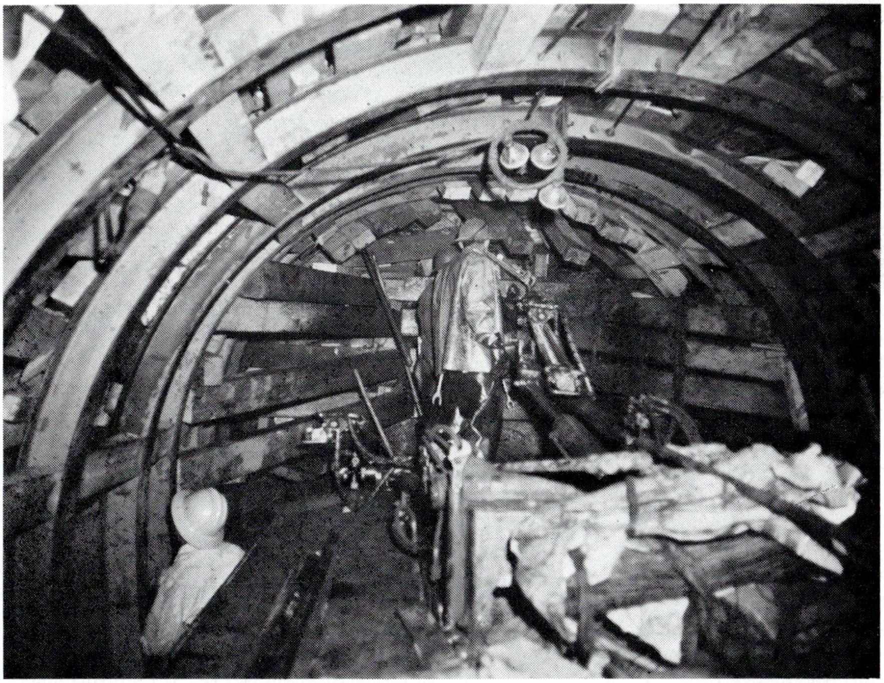

Fig. 113. Wood-plank spiling and breastboards used to contain and penetrate closely sheared squeezing ground. Notice timber spreaders and steel tie-bolts between steel ribs. Work at heading is conducted from a drill jumbo. Roberts Tunnel, Colorado. (Photo by Denver Board of Water Commissioners.)

Center line. The line at the center of the finished concrete lining of circular cross-section or a mole-bored tunnel.

Cherry picker. A swing-arm hoist used for switching muck cars.

Collar brace. Wood or steel brace placed between or on top of steel sets to maintain alignment.

Competent rock. Rock that stands without support in underground openings.

Crown. The highest portion of a tunnel. Also called the "roof" or "back".

Crown bar. Steel rod or rail or wood plank placed on top of or suspended below steel set or sets nearest heading and projecting forward to tunnel heading. Provides temporary support in arch until additional steel sets and lagging can be installed closer to the heading.

Double-jack. Hand drilling utilizing three men, two wielding hammers and a third holding and rotating a steel drill.

Drift. A small tunnel driven ahead of the main tunnel bore.

False set. A steel or timber set placed temporarily to expedite a tunneling operation, as, for example, a spiling operation.

Feeler hole. Hole drilled from tunnel heading or elsewhere to test geologic conditions beyond heading or in rocks adjacent to an underground opening.

Footblock. Wood block serving as a support at the base of a steel rib. Block on which ribs are erected.

Forepoling. Driving of wood or steel members into soft or incompetent rock in the arch of the tunnel to provide roof support beyond the tunnel heading. *See* Spiling.

French drains. Excavated channels in the tunnel invert filled with porous gravel and designed to control water flows especially during placement of concrete tunnel linings.

Gad. A sharp-pointed steel drill used to dig small holes in rock and to remove fractured rock.

Grout. Mixture of cement, water, and for special purposes, various additives. Pumped into fractured rock to fill voids, seal off water flows, and increase strength.

Gunite. Mixture of water, sand, and cement sprayed on exposed rock to provide a protective coating.

Heading. The working face at the limit of penetration of the tunnel.

Incompetent rock. Rock that will not stand in underground openings without support.

Invert. The portion of the tunnel below the center line. The bottom or floor of a tunnel.

Invert section. The section of the tunnel below the center line or spring line.

Jumbo. The movable platform which supports the drilling equipment used at the tunnel heading (Fig. 113).

Jump set. Steel or timber set placed between existing sets to provide additional support.

Lagging. Wood planking or blocks placed above and on sides of steel ribs to contain broken rock and to prevent its fall into an underground opening.

Left rib. The steel rib on the left side of the tunnel as one faces the tunnel heading.

Mole. A tunneling machine which cuts and abrades a tunnel of circular cross-section.

Muck. Disaggregated rock formed by blasting operation or by the free flow of incompetent material into an underground opening.

Mucking machine. The machine used to load broken rock for haulage out of underground excavations.

Overbreak. Rock removed from outside of pay line as defined in the specifications. Usually expressed as the per cent of material removed in a cross-sectional area relative to the area within the pay line. Also may be expressed as a volume per cent.

Panning. Metal sheets located so as to divert underground water flows, especially during placement of concrete lining.

Payline. The line at the periphery of a tunnel which defines the limits of tunnel section for which payment will be made for excavation under the terms of a contract.

Permanent support. The concrete or shotcrete lining of the tunnel. Steel and timber supports generally are regarded as temporary supports.

Pilot bore. A small tunnel opening driven in advance of the main, larger bore.

Popping rock. Stressed rock that fails with explosive violence and ejects large and small rock fragments with high velocity.

Portal. The entrance to a tunnel. Commonly located in weathered rock or unconsolidated deposits requiring special provisions for protection and support of the tunnel entrance and approach.

Progress. A measure of the speed and efficiency with which construction of any kind proceeds.

Rake. A timber or steel member used as a support and placed at an angle.

Rib. A curved steel segment comprising half of a steel set or support.

Right rib. The steel rib on the right side of the tunnel as one faces the tunnel heading.

Rock bolt. See Roof bolt.

Rockburst. A sudden explosive detachment of a rock slab or rock debris from the arch, wall, or invert of a tunnel.

Rockcrete. A nearly dry mixture of sand, crushed rock, and cement sprayed on the arch and walls of a tunnel to provide support.

Roof. The upper portion of a tunnel, especially the portion of the tunnel immediately above the center line. Also called the "back" or "crown".

Roof bolt. Steel bolt split at one end or with an expansible head inserted into drilled hole in rock to support the rock. Head of bolt is forced against a steel plate next to the rock or expansible head is spread by a screw action (Fig. 107).

Running ground. Clayey or sandy aggregate, usually water saturated, that flows freely into underground excavation.

Single-jack. Method of hand drilling with one man using a hammer and turning the steel drill.

Spalling rock. A rock mass under stress that yields thin slabs or wedges of rock by rapid failure. Slabs commonly form parallel to walls or arch of an opening rock.

Spiling. Pointed steel rods, rails, channel iron, or wood planks (Fig. 113) or logs driven into the tunnel heading to support incompetent rock until supports can be placed. Usually the spiling is driven above and beyond the supports nearest the heading or on top of and beyond a false set.

Spitting rock. A rock mass under stress that breaks and ejects small fragments with considerable velocity.

Spreader. Wood section placed between ribs to maintain tension on tie rods or preserve alignment (Fig. 113).

Spring line. The line across or parallel to the tunnel above which the steel ribs are bent to a radius of curvature different from that below it. Longest horizontal dimension between stee ribs. May or may not coincide with center line.

Squeezing ground. Incompetent material, generally clayey, that behaves plastically under stress, and tends to close the tunnel opening because of adjacent or superjacent loads.

Steel rib. One of the segments of bent structural steel bolted to another rib to form a steel support or steel set in the tunnel.

Steel set. Steel support consisting of two ribs bolted or welded together. May include a strut across the invert of the tunnel.

Stringer. In mining, a narrow vein. Also a longitudinal construction member in tunnel supports.

Strut. Steel or timber segment placed between steel ribs across tunnel invert to provide additional strength. Bolted or welded to bottom extensions of ribs.

Support. Any fabricated structure, steel, wood, or concrete, placed to prevent failure of rocks around underground openings.

Swelling ground. Rock that swells after being exposed. Usually increases in volume because of hydration of clay minerals in an altered rock.

Temporary support. Support placed to support tunnel rocks until permanent supports can be installed. All supports except the lining generally are regarded as temporary.

Tie-rod. Steel rod threaded at both ends (Fig. 113). Used as a connection between adjacent steel sets to hold sets in place and provide strength in the direction of the tunnel.

Timber lagging. Timber planks placed on sides and on top of steel sets to contain exposed rocks between the sets and to transmit loads to the steel sets.

Top heading. A small tunnel opening ahead of the full-size opening. Used to probe and place temporary supports in exceedingly incompetent or wet sections of tunnel.

Tunnel face. The tunnel heading.

Tunnel section. Outline of tunnel as measured at right angles to center line. Also, any portion of the tunnel measured parallel to the direction of the tunnel.

Tunnel supports. Wood, steel, or concrete structures placed to prevent collapse or failure of tunnel rocks.

Tunnel machine. A device rotating on a horizontal axis with cutting edges which enable excavation of a tunnel of circular cross-section. Tunneling machines either produce a peripheral circular cut confining subsequent excavation within the limits of the cut or abrade and remove all of the rock in the tunnel heading.

Wall plate. A longitudinal support, usually steel, which supports arch ribs along the walls of a tunnel.

Wedge. Wood wedge used to hold lagging or steel supports in place.

Weeper (Weep). A pipe or drilled hole in rock or concrete designed to relieve pressure of underground water on the tunnel lining.

REFERENCES

Proctor, R.J. and White, T.L., 1968. *Rock Tunneling With Steel Supports.* The Commercial Shearing and Stamping Co., Youngstown, Ohio, 271 pp.
Sandström, G.E., 1963. *Tunnels.* Holt, Rinehart, and Winston, New York, N.Y., 427 pp.
Szechy, K., 1967. *The Art of Tunneling.* Akadémiai Kiadó, Budapest, 891 pp.

Chapter 10

GEOLOGICAL INVESTIGATIONS DURING TUNNEL CONSTRUCTION

INTRODUCTION

The responsibilities of the geologist have not ended when construction of a tunnel is started. Now the emphasis is changed to day-to-day inspection and mapping of the tunnel as the heading advances and careful correlation of underground geological features with those mapped on the surface during feasibility and design stage investigations. Information gathered by geological mapping of the tunnel is used to fill in the details of the generally incomplete picture of the geology obtained by surface exploration, and in many situations, will increase the ability of the geologist to predict changing conditions beyond the tunnel heading through an increase in his understanding of the significance of features noted at the surface.

In some situations data derived from geological study of a tunnel may assist greatly in interpretation of poorly or incompletely understood surface features, and a reexamination of the surface geology over the tunnel in the light of the new data may be very fruitful.

A function of the geologist during construction of a tunnel is preservation of the safety of the bore as it is related to changing geologic conditions, and to contribute, to the extent possible, to the avoidance of emergencies created by sudden and unexpected intersection of highly incompetent rocks or heavy water flows.

GEOLOGICAL MAPPING OF TUNNELS

Geological mapping of tunnels and other underground openings has reached an advanced state of elegance and refinement, largely because of the efforts of mining geologists who are concerned with the geological details of economic mineral deposits. The techniques of the mining geologists are equally applicable to mapping of tunnels and irregular underground openings of all kinds (McKinstry, 1948; Forrester, 1946).

Geological maps of tunnels are made on appropriate scales, depending on the complexity of the geology. Commonly used scales are 1 inch = 20 ft. and 1 inch = 40 ft. or 1 inch = 50 ft. In a typical mapping operation the geology is plotted in the horizontal, or nearly horizontal, plane parallel to the center line and at a convenient elevation in the tunnel. For tunnels of relatively small size, say up to 15 ft. in diameter, the plane of mapping usually is at *springline elevation.* Springline elevation is the elevation of the widest portion of the tunnel. In tunnels supported with horseshoe-shaped steel ribs springline is at the bottom of the arch portion of the ribs. In circular tunnels springline has the same elevation as the center line.

Routine procedure includes identification of rock types and alterations and plotting strikes, dips, or plunges of all planar and linear rock elements. Side notes and/or symbols are used to characterize spacings, widths, and special aspects of the geology such as secondary mineralization and gouge in faults. The object of the

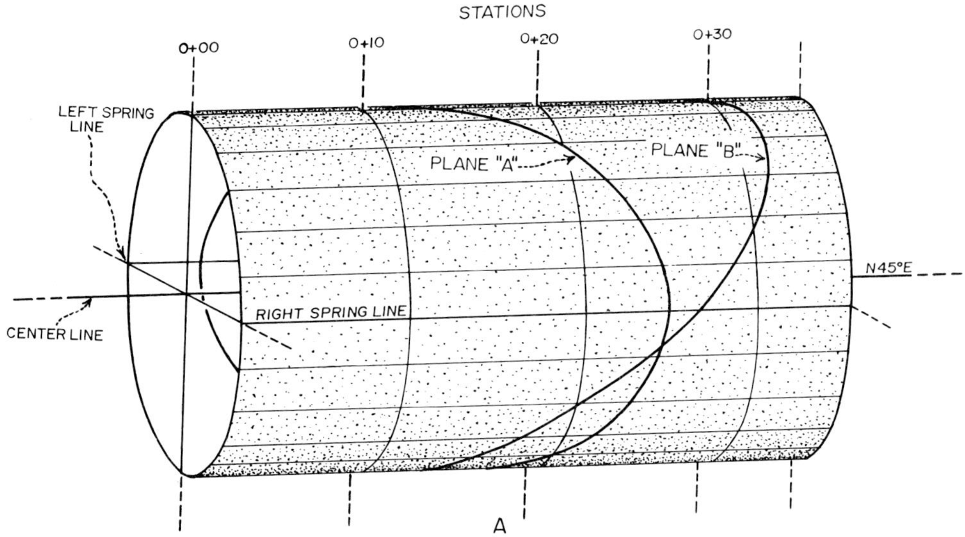

A

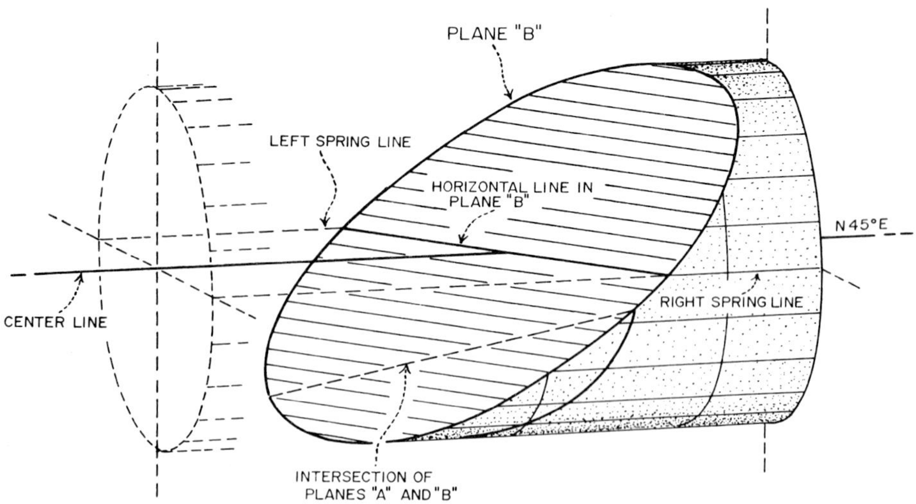

B

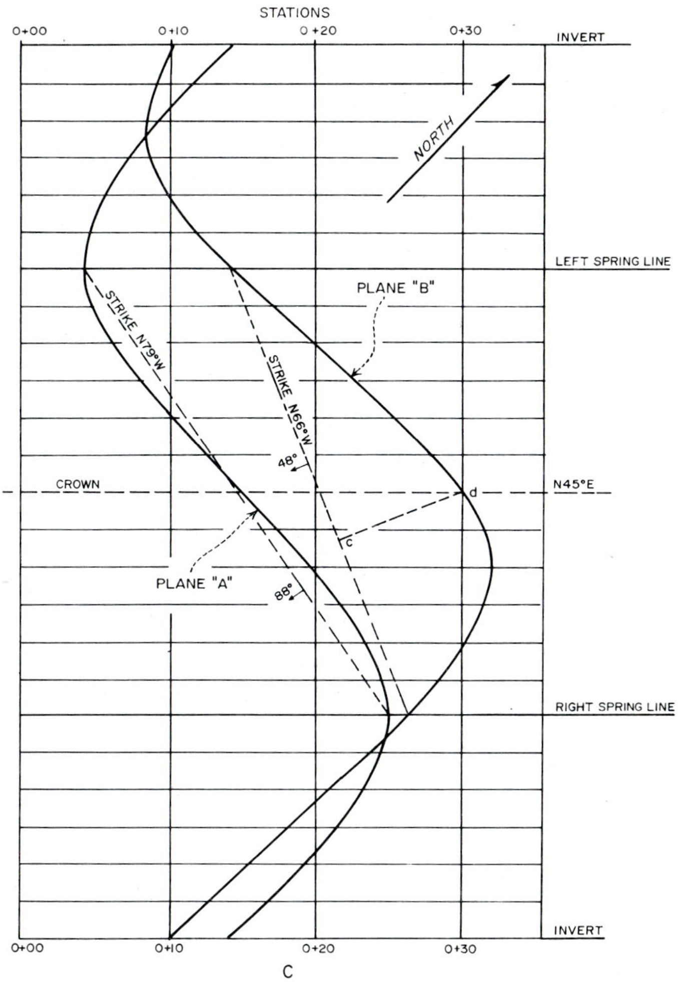

Fig. 114. A. Clinographic projection of a segment of a circular tunnel. Two planes *A* and *B* are intersected by the tunnel. Horizontal lines parallel to tunnel center line 15 degrees of arc apart. B. Plane *B* in Fig. 114A produces an elliptical trace in the tunnel opening. C. Cylinder in Fig. 114A unrolled to provide a grid for geological mapping in the tunnel. Traces of planes *A* and *B* are shown as they would appear if mapped on the grid.

mapping is to provide a complete graphical and written record of all features of geological significance and may require projection into the plane of the map of structures observed in the arch or invert. When planar linear elements do not intersect the plane of the map, sketches and photographs are included as part of the record. In any event, photographs, preferably in color, should be made at frequent intervals along the tunnel, particularly where there is a change in lithology and in the primary or secondary fabric.

Tunnels excavated by conventional methods (drilling and blasting) commonly have irregular outlines, and blasting emphasizes the presence of fractures and other planar fabrics such as bedding in sedimentary rocks and foliation in metamorphic rocks. Dips are readily measured with a clinometer, and strikes are determined by establishing the directions of horizontal lines in the planar fabric elements relative to the tunnel geometry and the direction of the center line. Measurement of strikes with a compass generally is not satisfactory because of distortion of the magnetic field by metal conduits, steel ribs, and rails, and electric power cables. Instead, directions may be measured with a protractor with a swing-arm index which enables determination of the angle between a strike direction and the known direction of the tunnel center line.

Circular tunnels excavated with moles present several problems to the geologist attempting to prepare a geologic map as the tunnel heading progresses (Cooper, 1968; Jack, 1969; Proctor, 1971). The generally rapid advance of the tunnel heading exposes rocks at a much more rapid rate than is usual in conventional excavation and requires more time spent in mapping each day. In addition, the bore often is covered with mud, dust, or rockcrete making detailed observation of geologic features very difficult or impossible. Planar or linear elements of fabric, when they can be seen, are not emphasized by blasting as in conventional mining, and appear on the smooth walls of the tunnel only as traces of planes or as points.

Under these circumstances geologic mapping should be done as soon as exposures are available for inspection and before rockcrete is applied. Extensive use of photography or videotape recording becomes essential, if a significant record is to be kept, and it may be necessary for the geologist or someone trained by him to plot the geology as a nearly continuous operation keeping pace with the advance of the tunneling machine.

The circular cross-section of a mole-bored tunnel enables easy preparation of a base for plotting geologic features as they are observed in a tunnel. Fig. 114A shows a segment of a tunnel of circular cross-section viewed from outside the tunnel in clinographic projection. Horizontal lines parallel to the center-line are drawn at intervals corresponding to 15 degrees of arc, and stationing along the tunnel is indicated by circles at right angles to the center line. The traces of two planes (*A* and *B*) on the cylinder defining the tunnel opening also are indicated. These planes might be taken to represent any planar geological feature such as a fault, a strong joint, a bedding plane, or a plane of foliation.

Fig. 114B illustrates the relationship of plane *B* to the geometrical elements of the tunnel bore and suggests that any plane intersecting the tunnel bore will provide a trace on the tunnel walls, crown, and invert that is a line (parallel to the center line), a circle (at right angles to the center line), or, in the general case, an ellipse.

The bearing relative to the north direction of any horizontal line in the plane is the strike of the plane.

Fig. 114C shows Fig. 114A unrolled to provide a base for geologic mapping and indicates the locations of the traces of planes *A* and *B* as they would appear on the base as viewed from above the tunnel. If the grid shown in Fig. 115C is plotted on a translucent plastic sheet, it may be used to map tunnel geology either on the assumption that the observer is outside of the tunnel or that he is inside the tunnel looking at its periphery.

Fig. 114C shows constructions for determining the strikes and dips of the two planes. Lines connecting intersections of the planes with right and left springlines yield the strikes. The dip is calculated by drawing a normal from the intersection of a plane with the crown perpendicular to the strike (line *c–d*). The elevations at the crown and springline are known, so that measurement of *cd* enables calculation of the dip angle by a tangent relationship.

When steel ribs or liner plates are installed as supports in a circular tunnel, convenient reference points of known locations are available around the periphery of the tunnel, depending on the design of the supports. The locations of tie bolts or spreaders in supports are very useful in establishing a grid of reference points or lines for constructing a base grid for geologic mapping.

Fig. 115 indicates the method for constructing a base grid different from that in Fig. 114 and, in some respects, more useful. The elements of the grid parallel to the

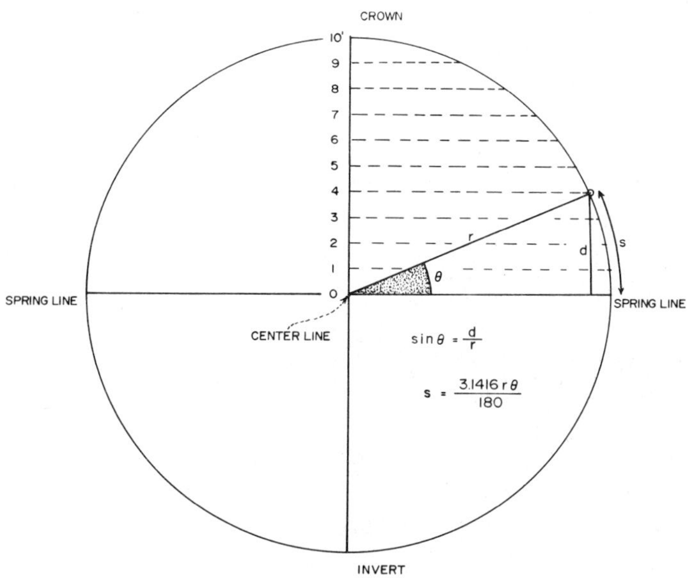

Fig. 115. Calculations for construction of the unrolled base grid for geological mapping shown in Fig. 116. Diameter of tunnel is 20 ft.

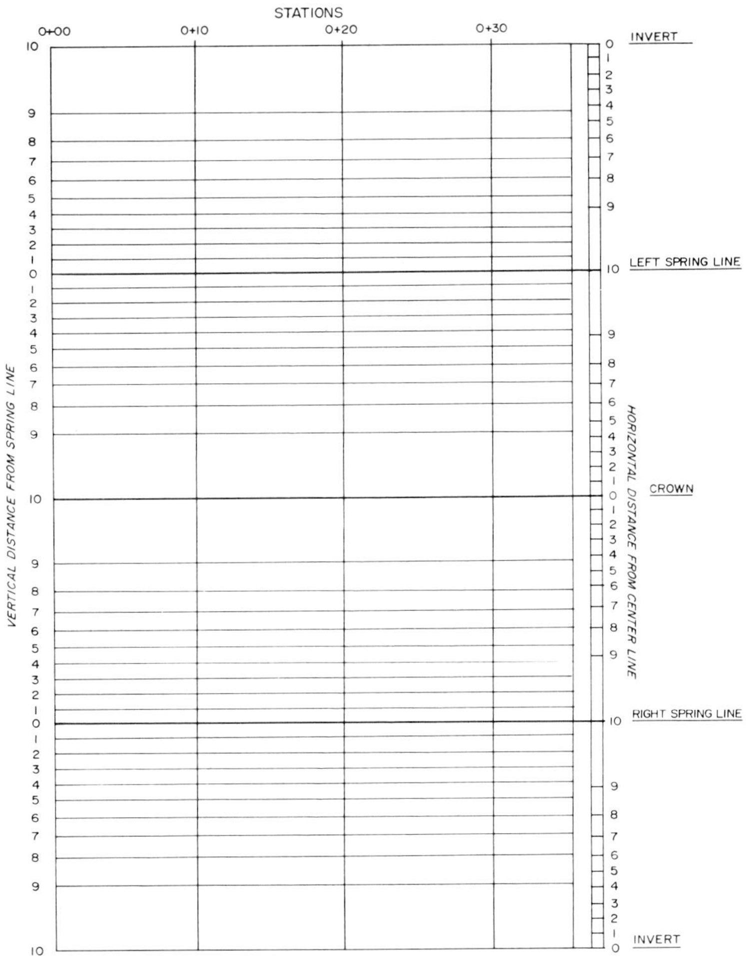

Fig. 116. Base grid for mapping geology in a circular tunnel of 20 ft. diameter. For construction details see Fig. 115.

tunnel center line are proportional to vertical distances below and above the plane of the springline and horizontal distances to the left and right of the center line in the plane of the springline.

Fig. 116 is a grid for a tunnel of 20-ft. diameter constructed by means of the calculations suggested in Fig. 115. Using the formulas included with Fig. 115 a grid is easily constructed for a circular tunnel of any radius.

Maps prepared with the grids in Fig. 114 and 116 may be rolled into cylinders which provide three-dimensional pictures of the tunnel geology as if it were viewed from outside of the tunnel. Transfer of data from the plots, usually in the office, to a horizontal plane at springline elevation and to a vertical plane through the center line yields a conventional geologic map and a longitudinal section. Similar grids may be constructed for mapping inclined openings or openings of noncircular but geometrical shapes, as in vaults or nonhorizontal diversion tunnels.

TUNNEL SUPPORTS AND THE GEOLOGIST

In tunnels excavated by conventional methods designs and specifications commonly include tunnel cross-sections for excavation in rock without support and cross-sections where supports are to be installed. If the tunnel is to be lined with concrete, it is obvious that the volume of concrete that must be placed is less in a smaller unsupported section than in a larger section, where room must be provided for supports around the periphery of the tunnel.

The decision as to whether a tunnel section is to be excavated as a "rock section" (unsupported) or a "steel section" (supported) must be made at the time that the heading is drilled and blasted. Subsequent enlargement of a rock section to provide for installation of supports, especially steel ribs, is time-consuming and costly, and should be avoided to the extent possible. The decision to install supports presents no problem in incompetent or highly incompetent ground, but requires a considerable exercise of judgment in marginal rock. Generally, for reasons of safety, light supports are routinely installed in marginal ground, although such ground may stand without failure for long periods of time.

In marginal rock a rock section may be supported with rock bolts, wire mesh (or metal mats), and rockcrete without greatly reducing the diameter of the bore, and, in many tunneling operations routine installation of such supports is common practice, and a steel section is excavated only where the rock is clearly incompetent.

The geologist plays an important part in estimating the competency of the rock and recommending the kinds and amounts of supports that should be installed. Generally, he makes his recommendation on the side of safety, but he must be cognizant of the fact that unnecessary installation of supports in competent or highly-competent rock represents an unwarranted expenditure of time and money.

Of particular concern to the geologist in his inspection and mapping of a tunnel is the possibility that some rocks behave competently on initial exposure in a tunnel heading but with time create problems. Examples are rocks that swell or slack over extended periods of time on exposure to ground water or moist tunnel atmosphere. Such rocks, although initially competent, may demonstrate long-term incompetency and can create problems far behind an advancing tunnel heading, or later, when the tunnel lining is completed.

Covering of rocks that have a tendency to swell with a layer of rockcrete or gunite generally is ineffective because slow penetration of the rocks by pore water under a pressure gradient and swelling may proceed in spite of a protective surface coating. The geologist, when he observes rocks with a swelling tendency, probably should advise against placement of a rockcrete or gunite coating so that the course of the swelling may be observed until the point has been reached where no more swelling is apparent. Pressures exerted by swelling rocks may be of a high order of magnitude and are capable of causing ultimate failure of tunnel lining and other kinds of confining supports, if an effort is made completely to contain the swelling rocks too early in the tunneling operation.

Rocks suspected of having a swelling tendency should be placed under observation in a container of water. If swelling is noted, samples should be sent to the laboratory for further examination and analysis.

In mole-bored tunnels the equipment generally is designed to excavate a circular opening of constant diameter, and no provision is made for "rock sections" and "steel sections" of different diameters. Supports in marginal rock are placed as soon as possible, and, if the supports do not prove to be adequate, remedial measures are taken as required at some distance behind the mole. When incompetent or highly incompetent rocks are encountered, the mole becomes ineffective in tunnel excavation, and it may become necessary to resort to conventional mining methods to advance the tunnel heading.

The responsibilities of the geologist with regard to installation of supports in mole operations consist chiefly of frequent inspection of the heading and, in addition, careful and constant examination of the tunnel behind the mole to determine whether supports might be required where they were not previously installed. Perhaps, as a partial responsibility, the geologist may recommend installation of rock mechanics instrumentation to make quantitative assessments of sections of the tunnel containing rocks that may fail over extended periods of time.

SPECIAL OPERATIONS

In addition to routinely preparing and maintaining current geologic maps and sections, the geologist has the responsibility of supervising or participating in a variety of special operations. Included are drilling of feeler holes or core holes ahead of the tunnel heading, estimation of the pressure, quantity, and distribution of ground water intersected in feeler holes, grouting operations to seal off water flows, determination of the chemical composition of ground-water flows intersected in the tunnel, measurement of rock and ground-water temperatures, and testing of the tunnel atmosphere for the presence of noxious or explosive gases, if this function is not delegated to a safety engineer.

Analysis of ground water may indicate the presence of solutes, especially sul-

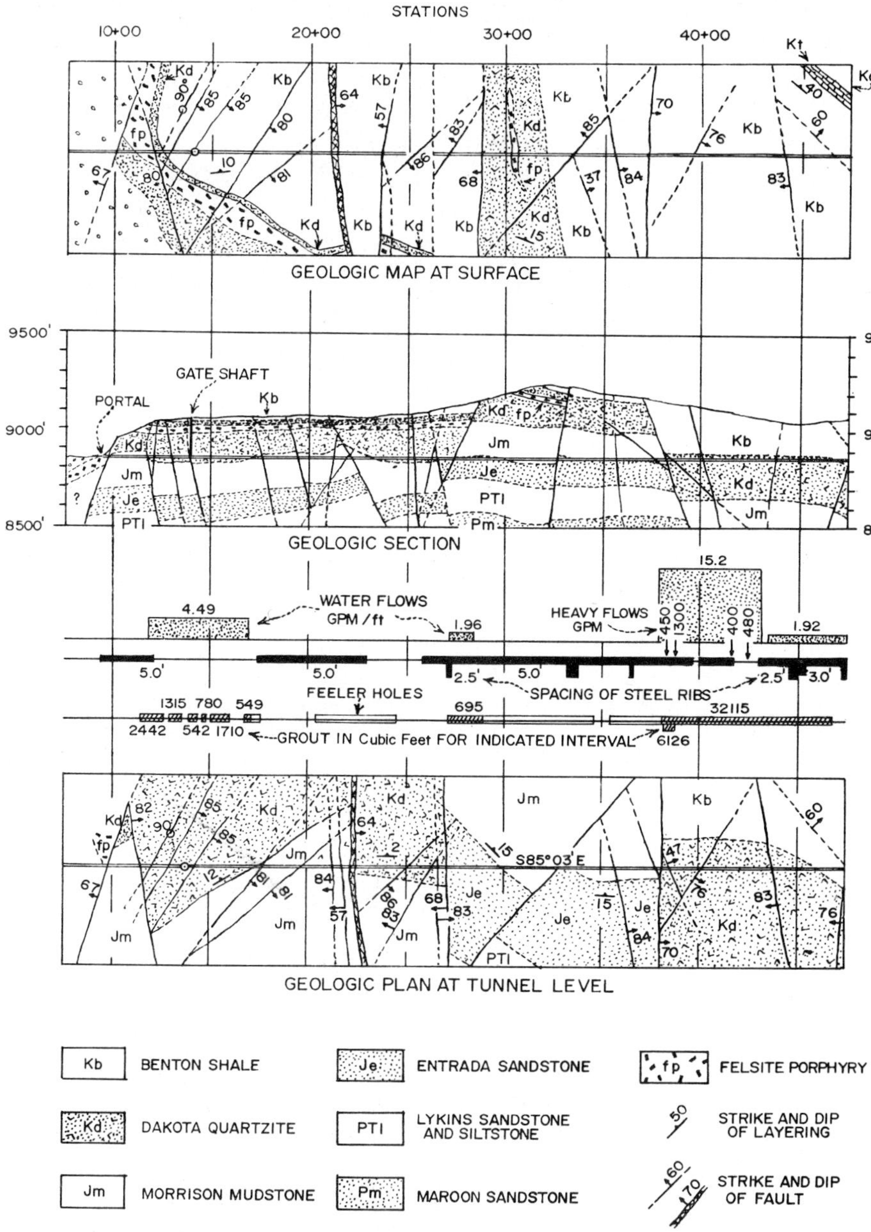

Fig. 117. Geologic plans and section with plots of data concerning ground water, supports, feeler holes, and grouting. West end of Harold D. Roberts Tunnel, Colorado.

fates, that react with concrete and may require special measures to prevent slow deterioration of the tunnel lining.

As a part of his responsibilities the geologist will find it very useful to maintain records of all special operations and to plot quantitative data from these operations in drawings showing plans and a section of the geology. This procedure emphasizes the significance of geological features as they are related to the need for feeler holes and grouting beyond the tunnel heading, and enables correlation of tunnel sections requiring light to heavy support with rock conditions.

A suggested procedure for maintaining a current record of special operations and correlating them with the geology is shown in Fig. 117, which shows the method used for plotting data from the Harold D. Roberts Tunnel in Colorado.

PROGRESS RELATED TO GEOLOGY

Progress measures the speed and economy with which a tunnel heading is advanced and is dependent on a variety of factors. Included are the method of tunneling and the efficiency of the engineering and labor forces, but, with these

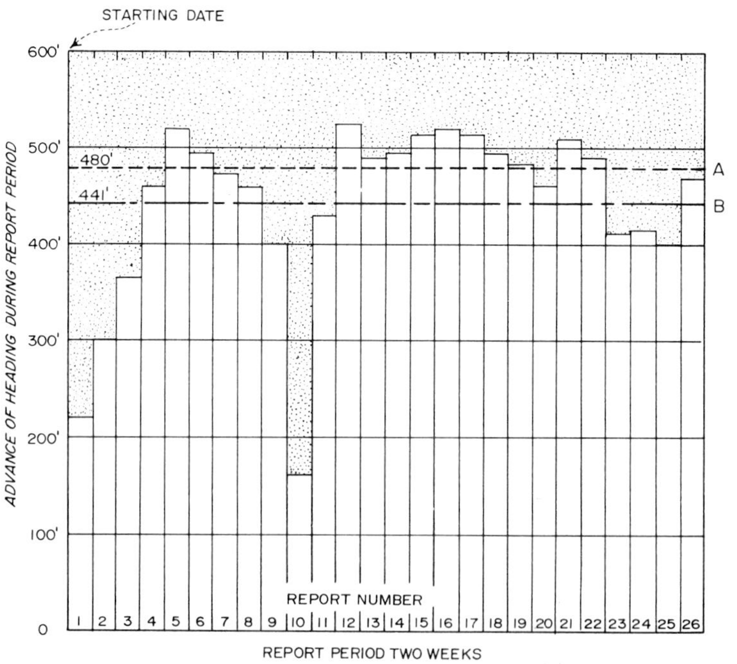

Fig. 118. Progress histogram. Scheduled average progress per report period (*A*) is 480 ft. Actual average progress per report at year's end is 441 ft. (*B*).

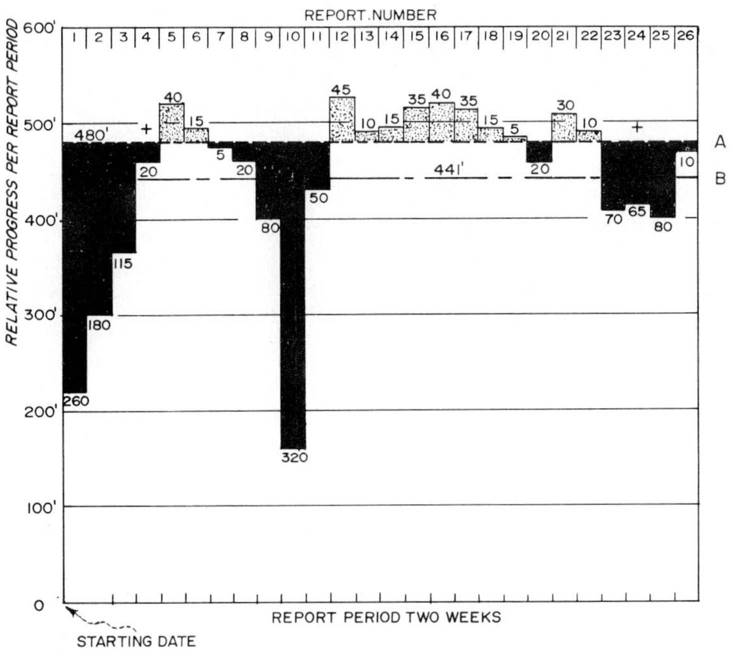

Fig. 119. Chart showing progress per report period relative to planned average (*A*). Actual average progress at year's end is 441 ft. (*B*).

being approximately constant, progress to a large extent is determined by the kinds of geological conditions that are encountered in the tunnel. Accordingly, a function of the geologist is a continuing study of the rate of safe advance of the tunnel heading as it is related to the geology. Aside from the interest associated with a particular tunneling operation, detailed study of progress, as it is related to geology, adds to the growing pool of data necessary for estimating costs during the feasibility and design stages of planning of other tunnels with similar geological conditions.

Several graphical methods for recording progress of a tunneling operation are available as indicated in Fig. 118–120. For all the plots in the illustrations an assumption is made that a tunneling operation using conventional mining methods has not been interrupted by labor difficulties or serious mechanical breakdown during a year's operation. It is also assumed that a summary engineering report of the operation is made every two weeks, and that geologic maps and sections are a part of the report, so that progress can be correlated with detailed geological conditions.

Fig. 118 illustrates a progress histogram. The average planned progress is 480 ft. per report period. An actual average progress of 441 ft. per report period was calculated at the end of a year's operation and plotted on the graph.

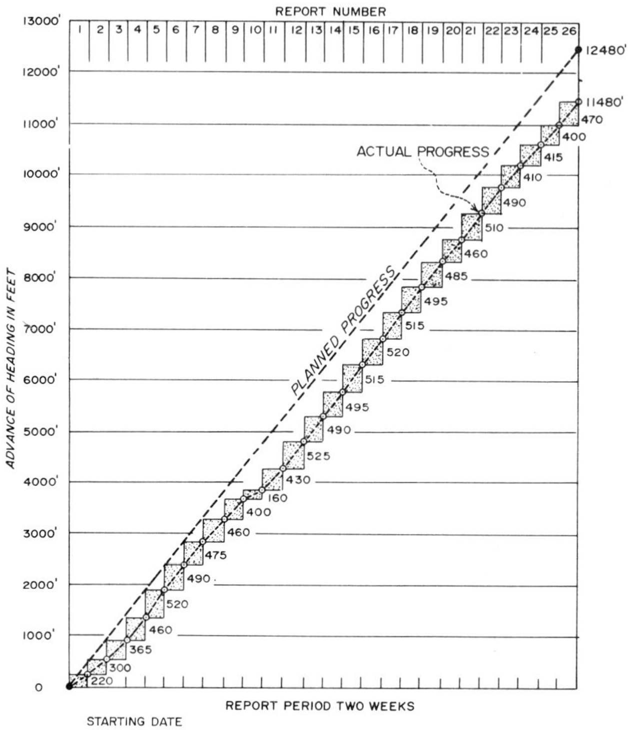

Fig. 120. Cumulative actual progress compared with planned progress. Based on same data that were used in constructing Fig. 118 and 119.

Fig. 119 utilizes the same data that were used to construct Fig. 118. Emphasis is placed, however, on the progress in feet per report period relative to an average planned progress and emphasizes intervals of the tunnel where progress is more or less than the planned average.

In Fig. 120 actual cumulative progress is compared with planned progress. This type of graph enables estimation of a trend in the actual progress to better advantage than the graphs in Fig. 118 and 119.

A particularly effective method for relating progress to geological conditions is indicated in Fig. 121. Progress per report period is compared with a planned average per report period and is correlated directly with the geology in longitudinal section and with special operations, such as grouting, required to advance the tunnel heading.

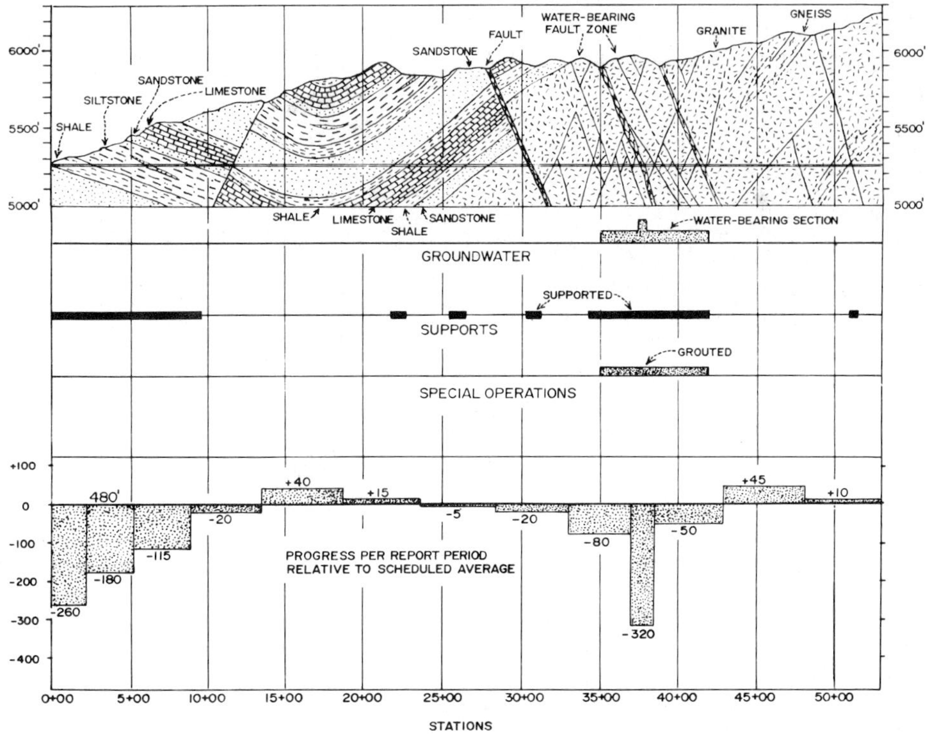

Fig. 121. Comparison of progress per report period with geological conditions as shown in a longitudinal section.

FINAL REPORT

The geologist is responsible for preparation of a final report in which he includes maps, sections, and drawings based on his continuing study of the tunnel during construction. Normally, this report is included with the final construction report for the tunnel and is a part of the permanent construction record.

REFERENCES

Cooper, K.R., 1968. A rapid, accurate method of mapping tunnels of circular cross section. *Bull. Assoc. Eng. Geologists,* 5: 63–68.

Forrester, J.D., 1946. *Principles of Field and Mining Geology.* Wiley, New York, N.Y., 647 pp.

Jack, H.A., 1969. Discussion: tunnel-mapping methods. *Bull. Assoc. Eng. Geologists,* 6: 151–156.

McKinstry, H.E., 1948. *Mining Geology.* Prentice-Hall, Englewood Cliffs, N.J., 680 pp.

Proctor, R.J., 1971. Mapping geological conditions in tunnels. *Bull. Assoc. Eng. Geologists,* 8: 1–43.

INDEX

Absorbed water, 110
Absorption of light, 10
Accordion fold, 89
Accumulates, definition, 9
Acicular fabric, 27
Adit, definition, 2
Adsorbed water, 110
Adsorption swelling, 111
Aerial photography, 163
AFMAG technique, 163
Agglomerate, volcanic, 36
Agricola, G., 7
Albite, 19
Alteration, associated with faults, 96
–, deuteric, 49
–, hydrothermal, 44
–, miscellaneous types, 50
–, of rocks, 39, 44
–, pneumatolytic, 49
–, rating, 186
Alternate tunnel alignments, 192
Aluminum hydroxides, 42
Amphibole, hornblende, 17
–, tremolite, 18
Andalusite, 20
Andesine, 19
Angle of shear, 62
Anhedral crystals, 26
Anhydrite, 16, 36, 185
–, swelling tendency, 123
Anisotropic fabrics, 67, 68
Anisotropic rocks, equal area plots, 174
Anisotropism, in rocks, 66, 85
Annular drainage, 143
Anorthite, 19
Anorthosite, 21, 22, 32
Anticline, asymmetrical, 89
–, symmetrical, 89
Aplite, 33, 43
Apparent dip chart, 150
Apparent resistivity, 159, 161, 162
Aquicludes, 113
Aquifers, 113
Arch, of tunnel, 218
Arching tendency, in tunnels, 106, 107, 108, 218
Argillization, 46
Arkose, 21, 22
Arkosic sandstone, 35
Asymmetrical anticline, 89
Asymmetrical syncline, 89
Augite, 17
Axis, of fold, 89
Axial plane, of fold, 89

Back, of tunnel, 218
Balmer, G., 58, 84
Barite, 16
Basalt, 32
–, porphyry, 32
Base grid for mapping, 225, 228
Batholith, 33
Bedding, in sedimentary rocks, 30
–, irregular, 30
–, regular, 30
Bedding-plane faults, 94
Bedrock maps, 148
Bentonite, 110, 123
Bents, 209
Biotite, 16
Biotite gneiss, 37
Biotite granite, 31
Biotite schist, 37
Black mica, 16
Blocking, in tunnels, 218
Blocky rock, 218
Body forces, 52, 54
Bonham-Carter, G., 179
Borehole deformation, measurement, 74
Bottom heading, 218
Bragg equation, 24
Brattice, 218
Breastboards, 218
Breccia, volcanic, 36
Breccia dikes, 96
Bronzite, 17
Bulkhead, 218
Bumps, in rocks, 63

Bytownite, 19

Calcarenite, 36
Calcite, 16, 48
Calcspar, 16
California switch, 218
Caldera, 98, 141
Canyons, as land forms, 138
Capillary water, 110, 111
Carbon dioxide, in rocks, 132
Carbonatization, 50
Caverns, in limestone, 119
Cement, in sedimentary rocks, 30
Center line, of tunnel, 219
Chain silicates, 11
Chalcedony, 20
Cherry picker, 219
Chert, 20
Chertification, 50
Chevron fold, 89
Chlorite, 15, 24, 42, 47
Chloritization, 47
Classification, grain size for crystalline rocks, 26
–, of geological materials, 182
–, of igneous rocks, 32
–, of land forms, 136
–, of metamorphic rocks, 37
–, of rock competency, 187
–, of sedimentary rocks, 35, 36
–, of tunnels, 4
–, rock minerals, 11
Clastic rock, definition, 28
Clastic rocks, size classification, 28
Clasts, definition, 28
–, shapes of, 29
Clay minerals, 15, 23, 42
Claystone, 35
Cleavage, in minerals, 12, 13
–, in rocks, 69
Cleavage folds, 89
Clinographic projection, 225, 226
Clinometer, 226
Clinozoisite, 19
Coarse tuff, 36
Coastline land forms, 139
Cockade fabric, 27
Coefficient, of internal friction, 59, 61
–, of mass fracture, 71
–, of permeability, 112
Cohesive strength, 58, 61
Collapse breccia, 97
Collar brace, 219
Colloform fabric, 27
Color, in minerals, 11
Color infrared photography, 165
Columnar sections, 151
Competency, classification, 187
–, final rating, 187, 188
Competency rating, 183
–, outline procedure, 184
Competent rock, 52, 219
Competent rocks, faulting, 93
Compressional seismic waves, 155
Compressional stress, 54
Compressive strength, 52
Computerization, of geologic data, 179
–, of tunnel costs, 193
Conchoidal fracture, in minerals, 13
Concordant igneous intrusions, 31, 34
Conductors, electrical, 159
Cone sheets, 33
Cones, volcanic, 34
Confidence rating, of geological projections, 194, 195
Conglomerate, 35
Conglomerate breccia, 35
Conical shear failure, 59
Conjugate shear fractures, 56
Construction record, 235
Conventional versus mole tunneling, 216
Cook, N.G.W., 65, 74, 84
Coon, R.F., 202
Cooper, K.R., 226, 235
Cordierite, 20
Core diameters, 165, 166
Core drilling, 165
–, at tunnel heading, 166
–, check list of procedures, 167
–, during feasibility studies, 191
–, during feasibility and design investigations, 166
–, logs, 168
Core holes, geophysical investigation, 172
–, in tunnels, 230
–, logs, 168
–, resistivity logs, 173
–, seismic velocity measurement, 173
–, water testing, 169
Cores, oriented, 170
Coulomb failure, 52
Coupled stress, 54
Creep, 52
–, steady state, 63
–, tertiary, 61
–, transient, 61

Critical distance, in seismic methods, 157
Cross-sections, geologic, 143, 150
Crosscut tunnel, definition, 2
Crown, of tunnel, 219
Crown bar, 219
Crushing strength, of rocks, 61
Cryptocrystalline fabric, 23
Crystallates, definition, 9
Crystalline dolostone, 36
Crystalline limestone, 36
Crystalline rocks, grain size classification, 26
Crystals, anhedral, 26
–, euhedral, 26
–, subhedral, 26
Cumulative progress, 234

Darcy, unit, 112
Darcy's law, 112
Dead-load weight, 57
Deere, D.U., 182, 202
Deformation, of elastic rocks, 87
–, by flow, 85
–, by fracturing, 85
–, of minerals under stress, 85
–, of plastic rocks, 88
–, of quasi-elastic rocks, 87
Dendritic drainage, 143
Depth of weathering, 40
Deshwar, K.H.S., 71, 84
Design stage studies, 144
– – –, geological, 181
Design-stage, geological investigations, 193
Detzlhofer, H., 131, 133
Deuteric alteration, 49
Diagenetic changes in rocks, 44
Diapirism, 137
Differential stress, 61
Differential thermal curves, 24, 25
Diopside, 17
Dip, 227
Dip needle, 158
Directed forces, 79
Discordant igneous bodies, 31
Dissolution, of limestone, 119
Dobrin, M.B., 155, 179
Dog-leg tunnel alignments, 192
Doloarenite, 36
Dolomite, 16, 48
Dolomite breccia, 36
Dolomite conglomerate, 36
Dolomitic marble, 37
Dolomitization, 48
Dolomitized limestones, 118
Dolostone, crystalline, 36
–, permeability, 118
Double-jack, 219
Drainage patterns, related to structure, 143
Drift, definition, 2, 219
Drill jumbo, 204
Ductile behavior, 52
Ductile material, 56
Duvall, W.I., 58, 65, 74, 84

Effective porosity, 111
Elastic material, 56
Elastic rebound, 80
Elastic rocks, deformation, 87
Elastic solid, 52
Elastic strain, in crystals, 87
Electrical configurations, in electrical prospecting, 160
Electrical methods, of geophysical investigation, 159
Electron microscope, 24
Elliptical overbreak, 106
Endogenous geomorphic processes, 135
Energy paths, in seismic methods, 156
Engineering factors, in competency ratings, 186
Enstatite, 17
Epeirogenic movements, 135
Epidote, 20
Epidotization, 50
Equal-area net, 174, 176
– – –, polar, 178
Equal-area plots, for anisotropic rocks, 174
– – –, in statistical analysis, 177
Equigranular-mosaic fabric, 27
Equigranular-mutual fabric, 27
Essential minerals, 31
Euhedral crystals, 26
Exogenous geomorphic processes, 135
Extension fractures, 56
Exterior boundary surface, 54
Extrusive igneous activity, associated landforms, 137
Extrusive igneous bodies, 31, 34

Fabric, 9, 66
–, acicular, 27
–, cockade, 27
–, colloform, 27
–, cryptocrystalline, 23
–, equigranular mosaic, 27
–, equigranular-mutual, 27
–, graphic, 27

Fabric, hypautomorphic-granular, 27
–, megacrystalline, 23
–, megascopic, 66
–, microcrystalline, 23
–, microscopic, 66
–, planar, 70
–, platy, 27
–, poikilitic, 27
–, porphyritic, 27
–, primary crystalline, 28
–, regional, 66
–, relict, 68
–, secondary crystalline, 28
Fabrics, anisotropic, 67, 68
–, from growth of crystals, 26
–, in rocks, 27
–, isotropic, 67, 68
–, of fragmental rocks, 28
False rock bursts, 82
False set, 219
Fan folds, 89
Fault filling, 93, 94
Fault gouge, swelling ground, 123
Fault scarp, 141
Fault-line scarp, 141
Fault zones, rating, 185
Faults, 92
–, associated landforms, 138
–, associated rock alteration, 96
–, associated topographic features, 141
–, bedding plane, 94
–, footwall, 92
–, gravity-slip, 98
–, hanging wall, 92
–, major, 98
–, normal, 92
–, related to rock competency, 93
–, reverse, 92
–, strike-slip, 92
–, transcurrent, 99
Feasibility studies, of tunnel locations, 144, 181, 191
Feeler holes, in tunnels, 214, 219, 230
Feldspathization, 50
Feldspathoid, leucite, 18
–, nepheline, 18
Felsite, 32
Ferromagnetic substances, 158
Fibrolite, 19
Fibrous fracture, in minerals, 13
Final construction report, 235
Final geological report, 235
Final heave, 124
Firmo-viscous material, 52
Fissility, 69
Flatjack method, for stress measurement, 74
Flexural strength, 55
Flexural-slip folds, 88
Flow deformation, 85
Flow fold, 89
Fluorite, 17
Fold anisotropy, in rocks, 85
Folded rocks, engineering aspects, 91
Folds, associated with faults, 90
–, axial plane, 89
–, axis, 89
–, flexural-slip, 88
–, hinge line, 89
–, in competent rocks, 104
–, in rocks, 88
–, isoclinal, 89
–, recumbent, 89
–, shapes, 89
–, shear, 88
–, slip, 88
–, subsidiary features, 90
Foliation, 36
Footblock, 219
Footwall, of fault, 92
Force, definition, 53
Forces, directed, 79
–, nonequilibrium, 77
–, surface, 53
–, unbalanced, 79
Forepoling, 219
Form energy, 26
Fossil water, 114
Fracture, conchoidal, 13
–, definition, 52
–, fibrous, 13
–, hackly, 13
–, in minerals, 13
–, irregular, 13
–, splintery, 13
–, uneven, 13
Fracture anisotropy, in rocks, 85
Fracture cleavage, 104
Fracture deformation, 85
Fracture permeability, 117
Fractured rocks, tunnel sections, 105
Fractures, in rocks, 55
–, in uniaxial stress, 56
Fragmental rocks, fabrics, 28
Framework silicates, 11
Free water, 110
French drains, 220

Gabbro, 32
Gad, 220
Garnet, 20
Garnetiferous mica schist, 21, 22
Garnetization, 50
Gases, in rocks, 109, 132
–, in tunnels, 230
Geologic cross-sections, 143, 150
Geologic maps, based on remote sensing, 147
– –, from ground observation, 147
– –, kinds, 147
– –, of surface bedrock exposures, 148
– –, of tunnel sites, 143
– –, symbols, 145
Geologic variables, in materials classification, 182
Geological investigations, data gathering, 152
– –, during design stage, 193
– –, during tunnel construction, 223
– –, feasibility studies, 144, 181, 191
– –, final design stage, 144, 181
– –, of tunnel locations, 135, 181
– –, reconnaissance, 140, 181
Geological mapping of tunnels, 223
Geological maps of tunnels, 223
Geological materials, classification, 182
Geological projection, uncertainties, 199
Geological report, final, 235
Geology, interpreted from geomorphology, 135
Geomorphology, endogenous processes, 135
–, exogenous processes, 135
–, related to subsurface geology, 135
Geophysical investigations, 154
Geophysical logging of core holes, 172
Geophysical methods, electrical, 159
– –, gravity, 158
– –, magnetic, 158
– –, outline summary, 154
– –, seismic, 155
Geothermal gradient, 116
Gibbsite, 42
Gilman mining district, Colorado, running ground, 128
Glaciation, development of land forms, 139
Glauconite, 24
Gliding, in crystals, 87
Gneiss, 37
Gouge, 93, 94, 96
–, photomicrographs, 95
Graben, 98, 141
Gradient, geothermal, 116
Grain-size classification, for crystalline rocks, 26
Granite, 31, 32
–, altered by weathering, 43
Granite porphyry, 21, 22, 32
Grant, F.S., 155, 179
Graphic fabric, 27
Gravimeter, 158
Gravity-flow tunnels, 4
Gravity methods, of geophysical investigations, 158
Gravity-slip fault, 98
Graywacke sandstone, 35
Great Appenine Tunnel, Italy–Switzerland, 132
Grid, for tunnel mapping, 225, 228
Ground water, in tunnels, 213
Ground-water geology, 109
Ground-water movement, in weathering, 42
Ground-water table, 113, 161
Grout, 213, 220
Grout mixtures, 215
Grouting, in tunnels, 230
–, of water-bearing zones, 213, 215
Gunite, 211, 220
Gypsum, mineral, 15
–, rock, 36, 185

Habit, in minerals, 13, 14
Hackly fracture, in minerals, 13
Halite, 16
Handin, J., 84
Hanging wall, of fault, 92
Harbaugh, J.W., 179
Hardness, indentation, 12
–, of minerals, 11
–, scale of, 11
–, scratch, 12
Heading, of a tunnel, 220
Heave, 124
Heft, definition, 14
Heinrich, E.W., 20, 37
Hematite, crystalline, 18
–, earthy, 15
Hinge, of fold, 89
Hinge line, of fold, 89
Histogram of progress, 232
History of tunneling, 1
Hogbacks, 137, 140
Hookean solid, 52
Hooke's law, 52
Hornback, V.Q., 200, 202
Hornblende, 17
Hornblende gneiss, 37
Hornblende schist, 37
Hornfels, 37

Horse-shoe section, 211
Horst, 141
Hydrogen sulfide in rocks, 132, 133
Hydrogeology, 109
Hydrostatic stress, 54
Hydrothermal alteration, 44
Hydrothermal solutions, in faults, 97
Hydroxyl, in minerals, 110
Hygroscopic water, 110
Hypautomorphic-granular fabric, 27
Hypersthene, 17

Igneous batholith, 33
Igneous bodies, concordant, 31
– –, discordant, 31
– –, extrusive, 31, 34
– –, intrusive, 31
Igneous intrusions, associated land forms, 137
– –, concordant, 34
Igneous laccolith, 34
Igneous rocks, 9, 30
– –, original joints, 102, 103
Igneous sills, 34
Igneous stock, 33
Ignimbrites, 30
Illite, 42, 46
Illite group, 24
Ilmenite, 18
Imagery, radar, 164
–, thermal, 164, 165
Inclined shaft, definition, 2
Incompetency, long-term, 229
Incompetent rock, 52, 220
– –, tunnels in, 205
Indentation hardness, 12
Induced permeability, 112
Infrared photography, 165
Infrared radiation, 164
Inselbergs, 139
Insulators, in electrical investigations, 159
Interior surface, 54
Internal friction, coefficient, 59, 61
Intrusion breccia, 93
Intrusive igneous bodies, 31
Invert, of tunnel, 220
Iron oxides, 42
Irregular fracture, in minerals, 13
Isobars, 52
Isoclinal fold, 89
Isopachs, 52
Isotropic fabrics, 67, 68
Isotropism in rocks, 66

Jack, H.A., 226, 235
Jaeger, J.C., 65, 74, 82, 84
Joint rose, 100, 101
Joint set, 99
Joint surfaces, characteristics, 101
Joint system, 99
Joints, 99
–, associated landforms, 138
–, equal-area plots, 100
–, in igneous rocks, 102, 103
–, in metamorphic rocks, 104, 105
–, in sedimentary rocks, 103
–, rating, 185
–, spacing, 101
–, star diagram, 100
–, surface characteristics, 101
Jumbo, 204, 220
Jump set, 220
Juvenile water, 114

Kahn, J.S., 179
Kaolin group, 23
Kaolinite, 42
Karst topography, 119
Karsts, 139
–, in limestone, 119, 120
Kelvin material, 52
Kink fold, 89
Koch, G.S., Jr., 179
Krumbein, W.C., 179
Kyanite, 18

Labradorite, 19
Laccolith, 34
Ladder drainage, 143
Lagging, 220
Lambe, J.W., 124, 133
Landforms, 136
–, along coastlines, 139
–, associated with igneous intrusions, 137
–, by erosion of layered rocks, 137
–, by landsliding, 138
–, by mass wasting, 138
–, by soil movement, 138
–, canyons, 138
–, classification, 136, 137
–, formed by extrusive igneous activity, 137
–, formed by glaciation, 139
–, related to faults and joints, 138
–, surfaces of low relief, 139
–, valleys, 138
Landslides, 142
Lapilli tuff, 36

Laterites, 41
Layered rocks, erosional forms, 137
Left rib, 220
Leucite, 18
Limestone, cavernous, 119
–, crystalline, 36
–, dissolution, 119
–, dolomitized, 118
–, micritic, 36
–, original porosity, 118
–, permeability, 118
Limestone breccia, 36
Limestone caverns, 119
Limestone conglomerate, 36
Limestone karsts, 119
Limestone sinkholes, 119, 120
Limonite, compact, 17
–, earthy, 15
Liner plates, steel, 210
Lining, in mole-bored tunnels, 212
–, in tunnels, 7, 208
–, in tunnels driven by conventional methods, 211
Link, R.F., 179
Liquid inclusions in minerals, 110
Liquid material, 56
Lithologic symbols, 149
Load pressures, 58
Loess, 139
Logging of core holes, instrumental, 172
Logs, form for plotting, 169
–, of core holes, 168
Longitudinal geological sections, 144, 196, 197
Longitudinal seismic waves, 155
Longitudinal tension fractures, 56
Long-term incompetency, 229
Long-wire method, 162
Lopolith, igneous, 34
Luster, adamantine, 10
–, glassy, 10
–, greasy, 10
–, metallic, 10
–, nonmetallic, 10
–, of minerals, 10
–, pearly, 10
–, resinous, 10
–, silky, 10
–, vitreous, 10

Magnesite, 17
Magnetic iron ore, 19
Magnetic methods of geophysical investigation, 158
Magnetic susceptibility, 158
Magnetite, 19
Magnetometer, 158
–, airborne, 158
Maps, bedrock, 148
–, from ground observation, 147
–, from remote sensing, 147
–, geologic, 143
–, geologic symbols, 145
–, of bedrock exposures, 148
Marble, 37
Marcasite, 19
Marker beds, 152
Marsh gas, 132
Mass fracture coefficient, 71
Mass wasting, associated landforms, 138
Matrix, in sedimentary rocks, 30
Mats, steel, 210
Maxwell material, 52
Mechanical properties of rocks, 54
Mechanical weathering, 41
Megacrystalline fabric, 23
Megascopic fabric, 66
Megascopic properties of minerals, 14
Meinzer unit, 112
Melilite, 17
Merritt, A.H., 202
Mesa, 140
Metamorphic rocks, 36
– –, definition, 9
– –, original joints, 104
– –, outline classification, 37
Metasomatism, 45
Meteoric water, 114
Methane, in rocks, 132
Methods of tunneling, 203
Mica, 15
Mica schist, 21, 22
Micritic limestone, 36
Microcline, 18
Microcrystalline fabric, 23
Microfracturing, 85
Microscopic fabric, 66
Migmatite, 37
Miller, R.L., 179
Mineral grains, deformation, 86
Minerals, absorption of light, 10
–, classification, 11
–, clay, 23
–, cleavage, 12, 13
–, color, 11
–, deformation under stress, 85
–, essential, 31
–, fracture, 13

Minerals, habits, 13, 14
–, hardness, 11
–, megascopic properties, 14
–, physical properties, 10
–, reflection of light, 10
–, resistance to weathering, 42
–, varietal accessories, 31
Modulus of elasticity, 52
Mohr construction, 60
Mohr envelopes, 60
Mohr's stress circles, 59, 60
Mohr's failure, 52
Mohr's theory of failure, 59
Mohs' scale of hardness, 11, 12
Moisture slaking, 111
Mole, 203, 220
Mole tunneling, 207
– –, advantages, 217
– –, limiting factors, 217
– –, versus conventional tunneling, 216
Montmorillonite, 42
Montmorillonite group, 24
Moorhouse, W.W., 20, 37
Muck, 220
Mucking machine, 220
Mudstone, 35
Multiple-stage tunneling, 204
Murray, D., 7
Muscovite, 15, 24, 46
Muscovite gneiss, 37
Muscovite schist, 37

Nappes, 98
Natural electrical currents, 162
Navajo Indian Irrigation Project, mole-bored tunnel, 207
– – – –, tunneling machine, 206
Necks, volcanic, 34
Nepheline (nephelite), 18
Newtonian material, 53
Nichols, T.C., 133
Nonequilibrium forces, in rocks, 77
Nonequilibrium slopes, 76
Nonsilicates, 10, 11
Normal faults, 92
Normal stress, 58

Obert, L., 58, 65, 74, 84
Obsidian, 32
Obsidian vitrophyre, 32
Oligoclase, 19
Olivine, 20
Opal, 18
Orientation of particles, in sedimentary rocks, 29
Oriented cores, 170
Original permeability, 112
Original porosity, 111
Orthoclase, 18
Orthosilicates, 11
Oso Tunnel, Colorado–New Mexico, 108
Overbreak, 220
–, elliptical, 106
–, in anisotropic rocks, 178
–, in tunnels, 72, 105
Overcoring method for stress measurement, 74

Pachecho Tunnel, California, 189
Panning, 220
Paragonite, 46
Parallel folds, 91
Paramagnetic substances, 158
Particle orientation, in clastic rocks, 29
Particle-size classification, pyroclastic rocks, 29
– – –, sedimentary rocks, 29
Payline, in tunnels, 72, 220
Peaked ribs, 209
Pegmatite, 32
Pellicular water, 110
Pericentral drainage, 143
Permanent support, 220
Permeability, 111, 112
–, coefficient of, 112
–, in limestones, 118
–, induced, 112
–, original, 112
–, primary, 112
–, related to fractures, 117
–, secondary, 112, 132
–, testing in rocks, 131
–, volcanic rocks, 113
Petrographic methods, 14
Petrographic thin sections, 21
– – –, photomicrographs, 22
Petrography, of unaltered rocks, 9
Phacoliths, 34
Photography, aerial, 163
–, in tunnels, 226
–, infrared, 165
Phyllite, 37
Physical properties, of minerals, 10
Piezometric level, 113
Pilot bore, 220
Pitchstone, 32

Pitchstone vitrophyre, 32
Plagioclase, calcic, 19
–, intermediate, 19
–, sodic, 19
Planar fabrics, 70
Plastic behavior, 53
Plastic material, 53, 56
Platy fabric, 27
Plunge of lineations, chart, 150
Pneumatolytic alteration, 49
Poikilitic fabric, 27
Poisson's ratio, 53
Pole, of plane, 176
Popping rock, 220
Pore water, 230
Porosity, 111
–, effective, 111
–, of sedimentary rocks, 111
–, original, 111
–, primary, 111
–, secondary, 111
–, total, 111
Porphyritic fabric, 27
Porphyry, 31
Portal, of a tunnel, 220
Potassium feldspar, 18
Press, F., 179
Pressure tunnels, 4
Pressures, of swelling ground, 123, 124
Primary permeability, 112
Primary porosity, 111
Proctor, R.J., 208, 222, 226, 235
Progress, cumulative, 234
–, graphical representation, 232, 233, 234
–, in tunneling, 220
–, related to geology, 232
Progress charts, 232, 233, 234
Progress histogram, 232
Projection, confidence rating, 194, 195
Projection, of geologic features, 195
Properties, of rock minerals, 14
Propylite, 46
Propylitization, 46
Protodyakonov, M.M., 69, 71, 84
Pyrite, 19
Pyritization, 46
Pyroclastic rocks, definition, 28
– –, size classification, 29
– –, swelling, 123
Pyroxene, augite, 17
–, bronzite, 17
–, diopside, 17
–, enstatite, 17
–, hypersthene, 17
Quartz, 20
Quartzite, 37
Quartzose sandstone, 35
Quasi-elastic rocks, deformation, 87

Radar imagery, 164
Radar shadows, 164
Radioactivity, measurement, 163
Rake, 220
Rebound, elastic, 80
Reciprocal thermal gradient, 116
Reconnaissance geological investigations, 143, 181
Recrystallization, of rocks, 87
Recumbent fold, 89
Reflection of light, 10
Refraction methods, geophysical, 155
Regional fabric, 66
Relict bedding, 68
Relict fabrics, 68
Relief, *see* Landforms
Remote sensing, 163
Residual stress, 80
Resistivity, apparent, 159, 161, 162
Resistivity logs, of core holes, 173
Resistivity methods, geophysical, 159
Reverse faults, 92, 141
Rhyolite porphyry, 32
Ribs, in tunnels, 220
Ribs, peaked, 209
–, steel, 209
Right rib, 220
Ring dikes, 33
Ring silicates, 11
Robbins, J.S., 203
Roberts Tunnel, Colorado, 72, 105
– – –, bulkhead, 215
– – –, case history, 199, 200
– – –, generalized section, 199
– – –, geologic plans and sections, 231, 232
– – –, highly competent rock, 188
– – –, incompetent shale, 190
– – –, numerical data, 201
– – –, running ground, 128, 129
– – –, squeezing ground, 125
– – –, surface exposures, 200
Robinson, C.S., 133, 200, 202
Rock alteration, 39
– –, associated with faults, 96
– –, at moderate depths, 44
Rock anisotropism, 66
Rock aquifers, 113
Rock bolts, 210, 220

Rock bursts, 62, 80, 220
– –, false, 82
Rock cleavage, 69
Rock competency rating, 183
– – –, outline procedure, 184
Rock fabric, definition, 9
– –, from growth of crystals, 26
Rock gases, 132
Rock isotropism, 66
Rock load pressures, 58
Rock mechanics, 51
Rock minerals, 10
– –, classification, 10
– –, megascopic properties, 14
Rock quality, 183
Rock quality designation, 182
Rock quality rating, 184
Rock salt, 16, 36
Rock section, in tunnels, 229
Rock type symbols, 149
Rockcrete, 220
Rocks, fold anisotropy, 85
–, fracture anisotropy, 85
–, igneous, 9, 30
–, mechanical properties, 54
–, metamorphic, 36
–, petrography, 9
–, sedimentary, 9, 35
–, squeezing, 120, 125
–, swelling, 120
–, weathered, 40
Roof, of a tunnel, 220
Roof bolt, 220
Rotation, of mineral grains, 87
Rotational stress, 54
RQD (Rock quality designation), 182, 183
Running ground, 128, 220
– –, in tunnels, 120, 130
– –, treatment, 130
Rupture, 53

Sand dunes, 139
Sandstone, 35
–, arkosic, 35
–, graywacke, 35
–, quartzose, 35
Sandström, G.E., 7, 208, 222
Sanidine, 18
Saprolite, 42
Saussuritization, 49
Scale of hardness, 11
Scarp, 141
Schist, 37
Schistosity, 36
Schmidt net, 174, 176
Scratch hardness, 12
Secondary permeability, 112
Secondary porosity, 111
Sections, columnar, 151
Sedimentary rocks, 9, 35
– –, bedding, 30
– –, classification, 35, 36
– –, marker beds, 152
– –, original joints, 103
– –, porosity, 111
– –, stratification, 30
– –, tension joints, 103
Seismic activity, 79
Seismic energy paths, 156
Seismic investigations, configurations, 173
– –, in bore holes, 174
Seismic methods, 155
Seismic velocities, 155, 156
– –, causes of variation, 173
– –, from core holes, 173
Seismic waves, 155
Selenite, 15
Self-grouting, 216
Self-potential method, 159
Semi-conductors, 159
Sensing, remote, 163
Sericite, 24
Sericitization, 46
Serpentine, 16, 49
Shaft, definition, 2
Shale, 35
Shear, 53
Shear angle, in deformation, 62
Shear failure, conical, 59
Shear folds, 88
Shear strength, 53
Shear stress, 53, 54, 58
Shear waves, 155
Sheet silicates, 11
Shotcrete, 211, 212
Siderite, 16
Side-looking radar imagery, 164
Silicates, 10
–, chain, 11
–, framework, 11
–, orthosilicates, 11
–, ring, 11
–, sheet, 11
Silicification, 45
Sillimanite, 19

Sills, igneous, 34
Siltstone, 35
Simplon Tunnel, Europe, 114
Single-jack, 221
Sinkholes, in limestone, 119, 120
Size classification, clastic rocks, 28
Slab breccia, 97
Slaking, 111
Slate, 37
Slickensides, 96
Slickline, 209
Slip, in faults, 92
Slip folds, 88
Slope analysis, 75
Slope failure, 76
Slope of equilibrium, 75
Slope retreat, 76
Slopes, equilibrium, 75
–, nonequilibrium, 76
Slump block, 142
Solution stoping, 97
Spalling rock, 221
Special operations, in tunnels, 230
Specific gravity, definition, 14
Spherical projection, 175
Spheroidal weathering, 43
Spiling, 221
Spitting ground, 81
Splintery fracture, in minerals, 13
Spreader, 221
Springline, in tunnel, 221
Springline elevation, 224
Springs, thermal, 115
Squeezing, rate, 127
Squeezing ground, 83, 120, 125, 221
– –, in tunnels, 127, 130
– –, materials, 126
– –, treatment, 130
Stagg, K.G., 84
Star diagram, for joints, 100
Starvation Tunnel, Central Utah Project, mole-bored tunnel, 209
Statistical analysis, 179
Staurolite, 20
Steady state strain, 62
Steel lagging, 210
Steel liner plates, 210
Steel mats, 210
Steel members, miscellaneous, 210
Steel rib, 209, 221
Steel section, in tunnel, 229
Steel set, 221
Stereographic net, 170
– –, use in graphical solutions, 171
Stereoplot, 172
Stock, igneous, 33
Stored water, 118
Straight Creek Tunnel, Colorado, 47
– – – –, swelling and squeezing ground, 189
Strain, 53
–, steady state, 62
–, transient, 62
Strain hardening, 55
Strain–time curves, 61
Stratification, 30
Stratigraphic sections, 152
Strength, cohesive, 58, 61
–, crushing, 61
–, flexural, 55
–, related to time, 62
–, temperature effects, 62
–, tensile, 53, 54
–, time-dependent, 62
–, triaxial compressive, 53, 55
–, ultimate, 61, 62
–, unconfined shear, 55
–, uniaxial compressive, 53, 54
Strength of rocks, from slope analysis, 75
Stress, compressional, 54
–, coupled, 54
–, definition, 54
–, hydrostatic, 54
–, normal, 58
–, residual, 80
–, shear, 58
–, tectonic, 80
–, tensional, 54
–, torsional, 54
–, uniaxial, 58
Stress concentrations, in tunnels, 64
Stress deformation of minerals, 85
Stress–strain curves, 56
Stress–strain data, 61
Stress–strain–time relationships, 55
Stress trajectories, 53
Stresses, tectonic, 74
Strike, 227
Strike and dip, from oriented cores, 170
Strike–slip faults, 92
Strike valley, 142
Stringer, 221
Strut, 221
Subhedral crystals, 26
Subsurface water, *see* Underground water
Sulfur dioxide in rocks, 132, 133

Supports, in incompetent rocks, 130
–, in mole-bored tunnels, 212
–, in tunnels, 7, 208, 221
–, in tunnels driven by conventional methods, 211
–, permanent, 220
Surface forces, 53
Susceptibility, magnetic, 158
Swell pressure, 124
Swelling ground, 120, 221
– –, anhydrite, 123
– –, fault gouge, 123
– –, hydrothermally altered rocks, 123
– –, in tunnels, 130
– –, materials, 122, 123
– –, mechanics and causes, 121
– –, pyroclastic rocks, 123
– –, treatment, 130
– –, weathered rocks, 123
Swelling pressures, 123
– –, of clay-rich rocks, 123, 124
Swelling rocks, mechanics and causes, 121
Swelling tendency, 230
Syenite, 32
Syenite porphyry, 32
Symbols, lithologic, 149
–, on geologic maps, 145
Symmetrical anticline, 89
Symmetrical syncline, 89
Syncline, asymmetrical, 89
–, symmetrical, 89
Széchy, K., 4, 7, 208, 222

Talc, 15, 49
Talobre, J.A., 84
Talus cone, 142
Tear faults, 99
Tectonic stress, 74, 80
Telluric currents, 162
Temperatures, effects on strength, 62
–, in rocks, 109
–, in tunnels, 230
–, of underground water, 114
Temporary support, 221
Tensile strength, 53, 54
Tension fractures, 56
Tension joints, sedimentary rocks, 103
Tensional stress, 54
Texture, of rocks, *see* Fabric
Thermal imagery, 164, 165
Thermal springs, 115
Thermal waters, 115
Thin sections, petrographic, 21
Tie bolts, 210
Tie-rod, 221
Timber lagging, 221
Timber members, 210
Time dependency of strength, 62
Time–distance curves, 157
Top heading, 221
Topographic features, *see* Landforms
Topography, *see* Landforms
Torsional stress, 54
Total porosity, 111
Tourmaline, 20
Tourmalinization, 50
Trachyte porphyry, 32
Transcurrent faults, 99
Transient strain, 62
Transverse seismic waves, 155
Trap, 32
Trapped water, 117
Trellis drainage, 143
Tremolite, 18
Triaxial compression, 57
Triaxial compressive strength, 53, 55
Triaxial stress, 57
Tuff, 36
–, coarse, 36
–, lapilli, 36
–, vitric, 36
Tunnel, adit, 2
–, crosscut, 2
–, definition, 2
–, drift, 2
Tunnel alignments, alternate, 192
– –, longitudinal sections, 196, 197
Tunnel atmosphere, 230
Tunnel construction, geological investigations, 223
Tunnel cross-section, in rock competency ratings, 186
Tunnel face, 221
Tunnel linings, 7, 208
Tunnel locations, avoiding canyons and valleys, 192
– –, geological investigations, 135, 181
Tunnel machine, 221
Tunnel mapping, grid, 225, 228
Tunnel portals, location, 191
Tunnel sections, 4, 5, 221
– –, in fractured rocks, 105
Tunnel supports, 7, 208, 221
– –, and the geologist, 229
Tunneling, access by smaller tunnels or shafts, 218

–, conventional methods, 204
–, history of, 1
–, in incompetent rock, 205
–, in running ground, 130
–, in squeezing ground, 130
–, in swelling ground, 130
–, multiple-stage, 204
Tunneling costs, computerization, 193
Tunneling machines, 203, 206
Tunneling methods, 203, 204
Tunneling terms, glossary, 218
Tunnels, arching tendency, 106, 107, 108
–, area of cross-section, 6
–, classification, 4
–, cross-sections, 4, 5
–, geological mapping, 223
–, gravity-flow, 4
–, ground water, 213
–, in running ground, 130
–, in squeezing ground, 130
–, in swelling ground, 130
–, overbreak, 72
–, payline, 72
–, pressure, 4
–, running ground, 121, 128
–, shapes in cross-section, 4, 5
–, squeezing ground, 120, 125
–, stress concentrations, 64
–, swelling ground, 120
–, uses, 4
–, volume of excavated rock, 6
Turner, F.J., 88, 108
Twinned crystals, 14
Twinning, in crystals, 14, 87

Ultimate strength, 61, 62
Unaltered rocks, 9
Unbalanced forces, 79
Unconfined shear strength, 55
Underground water, sources, 114
– –, sources of heat, 115
– –, temperatures, 114
Uneven fracture, in minerals, 13
Uniaxial compressive strength, 53, 54
Uniaxial stress, 58

Valleys, as landforms, 138
Varietal accessory minerals, 31
Vasquez Tunnel, Colorado, 48
– –, squeezing ground, 126
– –, swelling ground, 121
Velocities, of seismic impulses, 155, 156
Vermiculite, 24, 42
Vertical shaft, 2
Videotape recording, use in tunnels, 226
Viscous material, 53
Vitric tuff, 36
Vitrophyre, 32
Volcanic agglomerate, 36
Volcanic breccia, 36
Volcanic cones, 34
Volcanic necks, 34
Volcanic rocks, permeability, 113
Volcanic sink, 141
Voigt material, 53

Wahlström, E.E., 20, 37, 124, 131, 200, 202
Wall plate, 221
Warner, L.A., 200, 202
Water, absorbed, 110, 111
–, adsorbed, 110
–, capillary, 110, 117
–, chemically bound, 110
–, fossil, 114
–, free, 110
–, hygroscopic, 110
–, in rocks, 109
–, juvenile, 114
–, meteoric, 114
–, of percolation, 110
–, pellicular, 110
–, stored, 118
–, trapped, 117, 118
Water level, 113
Water table, 113
Water testing, in core holes, 169
Weathered rocks, engineering properties, 40
– –, swelling ground, 123
Weathering, 40
–, deep, 43
–, depth of, 40
–, leached environment, 42
–, mechanical, 41
–, related to ground-water movement, 42
–, related to mineral composition, 41
–, stagnant environment, 42
Wedge, 221
Weep, 221
Weeper, 221
Weiss, L.E., 88, 108
Wenner configuration, 160, 161
West, G.F., 155, 179
Wheby, F.T., 193, 202
White mica, 15
White, T.L., 208, 216, 222
Whitten, E.H.T., 88, 106, 108

Wire mesh, use in tunnels, 210
Work hardening, 128
Wrench faults, 99

X-ray diffractograms, 25
X-ray diffractometer, 24

Young's modulus, 53

Zeolites, 17, 49
Zeolitization, 50
Zienkiewicz, O.C., 84
Zoisite, 19